Marine Diesels

Marine Diesels

M. David Burghardt **George D. Kingsley**

United States Merchant Marine Academy
Kings Point, New York

Prentice-Hall, Inc., Englewood Cliffs, NJ 07632

Library of Congress Cataloging in Publication Data

BURGHARDT, M DAVID.
Marine diesels.

Includes index.
1. Marine diesel motors. I. Kingsley, George D., joint author. II. Title.
VM770.B776 623.8′7236 80-17164
ISBN 0-13-556985-0

Interior design and editorial/production supervision by Steven Bobker
Cover design by Edsal Enterprises
Manufacturing buyer: Joyce Levatino

Printed in the United States of America

10 9 8 7 6 5 4 3 2 1

Prentice-Hall International, Inc., *London*
Prentice-Hall of Australia Pty. Limited, *Sydney*
Prentice-Hall of Canada, Ltd., *Toronto*
Prentice-Hall of India Private Limited, *New Delhi*
Prentice-Hall of Japan, Inc., *Tokyo*
Prentice-Hall of Southeast Asia Pte. Ltd., *Singapore*
Whitehall Books Limited, Wellington, *New Zealand*

Contents

Preface

This book has been prepared to assist students in marine diesel engineering courses, particularly those courses emphasizing operating practice and maintenance procedures. Most of the material was derived from lecture notes used in courses in marine diesel engineering offered at the United States Merchant Marine Academy. We cover the essentials of design and construction without detailing actual engine specifications.

There is a growing interest in diesels as they provide a fuel-efficient alternative to the gasoline engine. The railroads and trucking industry rely on high speed diesels as their power source. The diesel engine plays and will increasingly play an important role in the transportation industry. This text has been developed and organized with the marine diesel engineer in mind. Nevertheless, since much of the material deals with fundamentals of diesel engines and with developing an understanding of them, the marine bias is limited.

The purpose of this text is to supply information for students and operational personnel in technical fields where operational, repair, and troubleshooting aspects of diesel engineering are emphasized. We do not aim to replace repair manuals or handbooks already available. These contain the necessary specifics on equipment design, installation, and repair. However, they do not serve well as texts where the desire is to create understanding and the transfer of knowledge. With understanding as a foundation, the information provided in the handbooks and manuals becomes more comprehensible and useful.

The authors are indebted to many people for their assistance, advice and criticism during the development of this text and wish to thank Mrs. Joan Feret for typing the final manuscript.

M. David Burghardt
George D. Kingsley

Marine Diesels

1

Engine Fundamentals and Thermodynamics

The diesel engine is one of the power sources that drive our economy. It is found in many situations and its applications range from automotive to marine, from small to medium to very large, from auxiliary applications to main power generating stations. This book will focus on marine applications of the diesel engine, but it should be remembered that often the same engine is used ashore, with minor modifications to some engine systems.

Total analysis of the diesel engine is quite complex, dealing with mechanical and thermal design and manufacturing processes. We will be interested in these in a qualitative sense, not in a quantitative sense. The best way to operate and service any device is to understand its design and design problems, although it is not necessary to be able to design the device. This is very true for diesel engines, and the intent of this book is to create an understanding of principles so operating procedure, maintenance practice, and troubleshooting evolve in a logical manner.

1.1 Cylinder Nomenclature

Let us analyze a cylinder with the various components found on and in it. Figure 1-1 illustrates the various components and volumes. The diameter of the cylinder is called the *bore*. When the piston is at the limit of its upward (or, in some instances, outward) travel, it is at *top dead center* (TDC); any further rotation of the engine will cause the piston to move down (or back in). When the piston is at its inward most point, it is at *bottom dead center* (BDC); any further engine rotation will cause the piston to move upward. The vertical distance that the piston travels between TDC and BDC is called the *stroke*. The *displaced volume* is

$$\text{Displaced volume} = \frac{\pi}{4}\,(\text{Bore})^2\;(\text{Stroke})$$

This is simply the formula for the volume of any cylinder; the bore is the diameter and the stroke is the altitude.

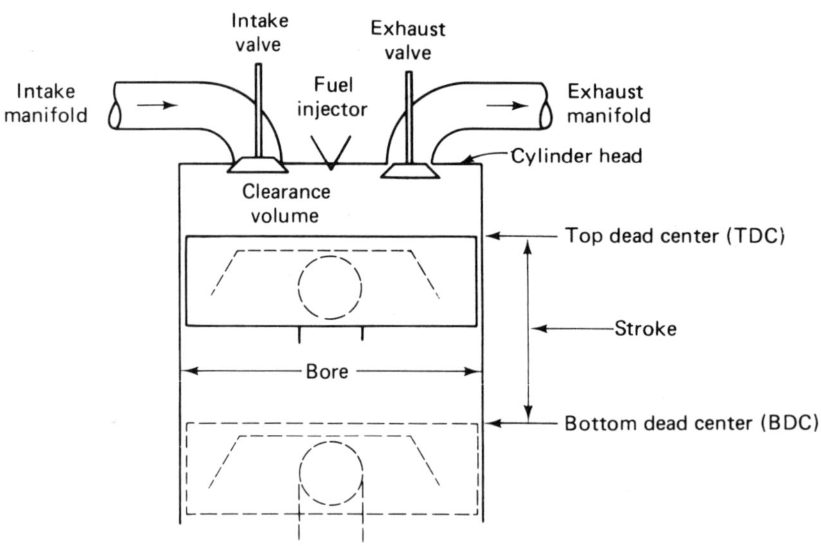

Figure 1-1. Schematic of major cylinder components and dimensions.

The piston cylinder size is always given as bore times stroke; for instance, 3 1/2 in. × 4 in. indicates a bore of 3 1/2 in. and a stroke of 4 in. In marine use, bores range from about 89 mm × 89 mm (3 1/2 in. × 3 1/2 in.) to 1050 mm × 1750 mm (41 1/3 in. × 68 7/8 in.). The *total engine displacement* is the sum of the displaced volume of all the cylinders.

The space between the piston at TDC and the cylinder head is the clearance volume. The cylinder head seals the top of the cylinder and typically has openings for intake, exhaust, and a fuel injector. (The specific head openings will be analyzed in Chapter 4.) Air enters the cylinder through the intake manifold and intake valve. The opening of the intake valve must be timed to allow the maximum amount of air to enter, and its closing must be timed so that no air escapes.

The exhaust valve functions in an analogous manner, opening to allow the products of combustion to leave the cylinder after they have transmitted energy to the piston, and allowing no leakage at other times.

Instead of having inlet and exhaust valves, the cylinder may be constructed with inlet and exhaust ports. Figure 1-2 is a schematic of a *ported cylinder*. The ports must be located at different levels. In this instance the exhaust port uncovers first and the intake port last, closing in reverse order. The use of ports eliminates the need for valves, but there are trade-offs in engine performance. These will be analyzed at various points in the succeeding chapters.

The function of the fuel injector is to supply an energy source to the engine, providing a metered amount of fuel to the cylinder at the appropriate time.

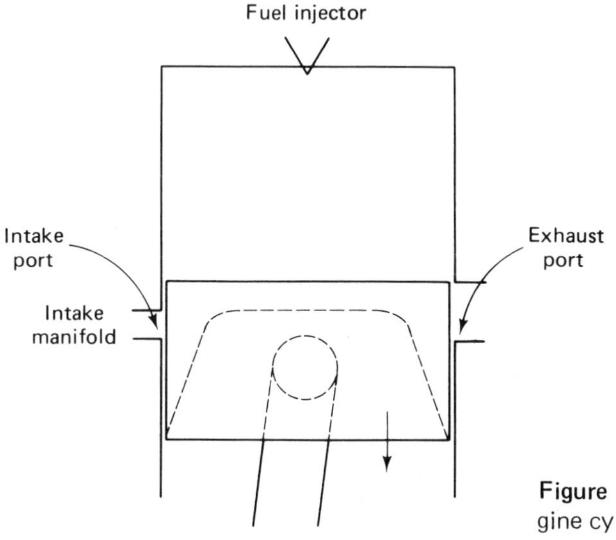

Figure 1-2. Schematic of a ported engine cylinder.

1.2 Thermodynamics

In any field dealing with energy, a knowledge of thermodynamics is important. Thermodynamics is the science that deals with energy, energy transformations, and changes of energy from one form to another. Diesel engine operation involves several such transformations. Consider the following energy flow

Chemical energy	$\longrightarrow$	Thermal energy	$\longrightarrow$	Mechanical energy	$\longrightarrow$	Shaft energy
Fuel supplied to engine	$\longrightarrow$	Combustion raises air/fuel temperature and pressure	$\longrightarrow$	Piston moves because of high pressure	$\longrightarrow$	Shaft rotates after translating piston motion

(1.1)

Chemical energy is received from the fuel injector. Combustion occurs, increasing the temperature and pressure of the trapped gases. The high gas pressure forces the piston downward, and its motion is converted by mechanical means into rotary motion, which is the form of the engine's energy output.

Two laws of thermodynamics apply in this situation. The First Law of thermodynamics states that energy is conserved. The Second Law of thermodynamics states that the "quality" of energy is not conserved. Quality in this instance refers to that part of the energy that can be used for work. Thus, looking at the energy flow of an engine, note that the total energy in must equal the total energy out, but the useful part (the work portion) of the energy will decrease with each energy transformation.

This leads to the concept of efficiency: how well the energy transformation occurred; actual versus ideal. Efficiency may be defined as

$$\eta = \frac{\text{Output}}{\text{Input}} = \frac{\text{Desired effect}}{\text{Cost of effect}}$$

The combustion process will never be ideal or complete; some of the fuel will not burn or will oxidize only partially, forming carbon monoxide, rather than carbon dioxide. Thus, the chemical energy of the fuel will not be totally utilized in raising the temperature and pressure of the air/fuel mixture. Combustion efficiency describes how well the combustion process occurred. The movement of the piston by the gas also will not be ideal, because of friction between the piston and the cylinder, leakage of the gas past the piston, and nonuniformities in gas pressure within the cylinder. Thus, the work the gas indicates it did on the piston and the amount of work the piston is able to transmit to the connecting rod are different. Furthermore, there are losses associated with the transformation of the reciprocating motion of the piston to the rotary motion of the shaft.

The efficiencies frequently encountered are the overall engine efficiency

$$\eta_{\text{overall}} = \frac{\text{Shaft work developed}}{\text{Chemical energy supplied}} \tag{1.2}$$

and the mechanical efficiency

$$\eta_m = \frac{\text{Shaft work developed}}{\text{Indicated work by gas}} \tag{1.3}$$

Equation *(1.2)* relates how efficiently the engine converted chemical energy into shaft work and equation *(1.3)* relates how efficiently the engine is mechanically designed.

The area that we will investigate in more depth is the conversion of thermal energy into mechanical energy, the conversion of the thermal energy of the gas into the mechanical energy (piston movement) of the engine. To accomplish this, a model of the diesel engine must be developed. This will enable prediction in advance of the effect of various parameters on engine performance.

Ideal Gas Law

Predicting engine behavior in advance requires determining how the substances in the cylinder (air, fuel, gases) will behave. Fortunately, for most engineering calculations the ideal gas law predicts the behavior of air and exhaust gas with sufficient accuracy. This law is written as

$$p = \rho RT \tag{1-4}$$

where p is the pressure, ρ is the density, R is the gas constant, and T is the absolute temperature.

The value of R depends upon the gas under consideration. The two gases of most concern are shown in Table 1-1 in both English and SI units.

TABLE 1-1

	R $\left(\frac{\text{ft-lb}_f}{\text{lb}_m\text{-}^\circ\text{R}}\right)$	R $\left(\frac{\text{kJ}}{\text{kg-}^\circ\text{K}}\right)$
Air	53.34	0.287
Exhaust	53.54	0.288

The units for the terms in equation *(1.4)* are

	English	S.I.
Pressure (p)	pounds force per square foot	kilopascals (kilonewtons per square meter)
Density (ρ)	pounds mass per cubic foot	kilograms per cubic meter
Temperature (T)	degrees Rankine	degrees kelvin

Thus, we may calculate the density of air at atmospheric pressure (14.7 psia, 101.3 kPa) and a temperature of 77°F (25°C).

$$p = (14.7 \text{ lb}_f/\text{in.}^2)(144 \text{ in.}^2/\text{ft}^2) = 2116.8 \text{ lb}_f/\text{ft}^2$$

$$T = 77 + 460 = 537°\text{R}$$

$$\rho = \frac{p}{RT} = \frac{(2116.8)}{(53.34)(537)} = 0.0739 \text{ lb}_m/\text{ft}^3$$

Air Standard Diesel Cycle

The air standard diesel cycle was developed to model engine behavior. This model represents a thermodynamic cycle, not a mechanical (or actual engine) cycle. Thus, we can analyze the cycle and determine the result of certain effects without having to get into the details and complications of actual engine performance. The development of a thermodynamic cycle entails several assumptions, the first of which is that the working substance has a constant mass and that the substance is air. This means that the air does not leave the cylinder but stays in the cylinder at all times. Instead of fuel being burned, it is assumed that there is a heat addition process; and instead of there being an exhaust process, it is assumed that heat is rejected. These assumptions allow a constant mass of air to pretend that it is operating on the diesel cycle. Figure 1-3 shows the air standard diesel cycle; the processes are noted to the right of it. The enclosed shaded area is equal to the net work. For each revolution of the crankshaft there is one complete air standard diesel cycle. It is important that the enclosed area is equal to the net work, for this allows us to calculate the work for the cycle and gives a reason for being interested in p-V diagrams for an actual engine.

Since the shaded area represents the work per revolution of the air standard diesel, determining that area is important. This task would be made easier if the work area were in the shape of a rectangle. Since there is no reason not to, we can reshape the area to that shown in Figure 1-4.

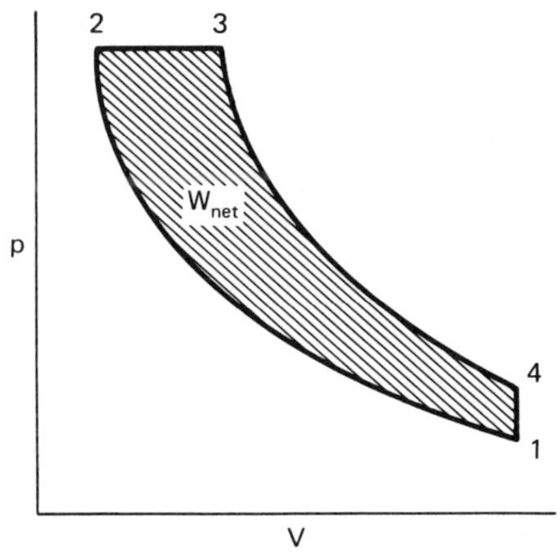

Thermodynamic processes

1 - 2 Adiabatic (q = 0) compression

2 - 3 Heat addition at constant pressure

3 - 4 Adiabatic expansion

4 - 1 Heat rejection at constant volume

Figure 1-3. Air standard diesel cycle.

The rectangular area will have a length equal to the piston displacement $(V_1 - V_2)$ and a height equal to the *mean effective pressure.* W_{net} is the same. Thus, the mean effective pressure is defined as

$$\text{mep} = \frac{W_{net}}{\text{Piston displacement}} \tag{1-5}$$

If we know the indicated mean effective pressure and the piston displacement, we can calculate the indicated work. Furthermore, if the power revolutions per unit time are known, the engine power may be

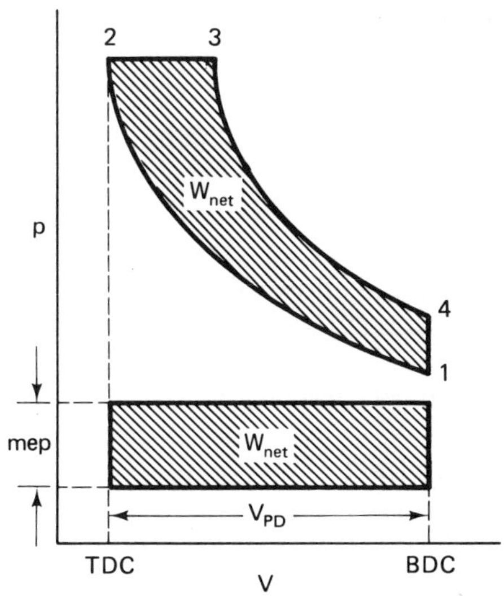

Figure 1-4. Mean effective pressure.

determined. For example, given the following information concerning an air standard diesel engine

$$\text{mep} = 200\ \text{lb}_f/\text{in.}^2 = 1379\ \text{kPa}$$

$$\text{Cylinder bore and stroke} = 12\ \text{in.} \times 18\ \text{in.} = 305\ \text{mm} \times 457\ \text{mm}$$

$$\text{Number of cylinders} = 8$$

$$\text{Power revolutions} = 400\ \text{rpm} = 6.667\ \text{rps}$$

$$W_{net} = (\text{mep})\ (\text{Piston displacement})$$

$$\text{Piston displacement} = (\text{Area})\ (\text{Stroke}) = \frac{\pi}{4} d^2 L$$

$$\frac{\pi}{4}\frac{(144)\ (18)}{(144)\ (12)} = \frac{3\pi}{8}\ \text{ft}^3 \qquad \frac{\pi}{4}\frac{(305)^2}{(1000)^2}\frac{(457)}{1000} = 0.0334\ \text{m}^3$$

$$W_{net} = (200)\ (144)^*\ (3\pi/8) \qquad W_{net} = (1379)\ (0.0334)$$

$$= 33{,}929\ \text{ft-lb}_f/\text{cycle} \qquad = 46.06\ \text{kJ/cycle}$$

Notice that $W_{net} = (\text{mep}) \times LA$ where A is the cross-sectional area. If we multiply W_{net} by the number of power revolutions, n, the engine power may be determined.

$$\text{Power} = nW_{net} = (\text{mep})\,LAn \tag{1.6}$$

Very often the mean effective pressure in equation *(1.6)* is written as p; thus the power is equal to $pLAn$. To obtain units of horsepower, a conventional power term in English units, it is necessary to divide $pLAn$ by 33,000 which is the number of ft-lb_f/min in one horsepower.

Continuing

$$\text{Power} = \frac{(33{,}929)\ (400)}{33{,}000} \qquad \text{Power} = (46.06)\ (6.667)$$

$$= 411.3\ \text{hp} \qquad = 307\ \text{kW}$$

The theoretical model for the diesel cycle allows the calculation of work for a variety of conditions and trends may be determined as these variables are changed.

The thermal efficiency for the air standard diesel cycle is

$$\eta = \frac{W_{net}}{Q_{in}} = \frac{\text{Desired effect}}{\text{Cost of effect}} \tag{1.7}$$

*144 is a conversion factor used in converting mep in $\text{lb}_f/\text{in.}^2$ to mep in lb_f/ft^2.

There are two important ratios which can be used in developing an expression for the thermal efficiency. The first is the compression ratio, r.

$$r = \frac{\text{Volume at BDC}}{\text{Volume at TDC}} = \frac{V_1}{V_2} \tag{1.8}$$

The second is the cut-off ratio, r_c

$$r_c = \frac{\text{Volume at end of heat addition}}{\text{Volume at start of heat addition}} = \frac{V_3}{V_2} \tag{1.9}$$

In an actual engine, the cut-off ratio refers to the volume ratio at the start of injection to that at the end of fuel injection. The thermal efficiency may then be reduced to the following form

$$\eta = 1 - \frac{1}{(r)^{k-1}} \left[\frac{r_c^k - 1}{k(r_c - 1)} \right] \tag{1.10}$$

where $k = 1.4$ for air.

Figure 1-5 illustrates the effect of the cut-off ratio on the overall thermal efficiency. As the cut-off ratio increases, there is more power. This means there is a longer period of heat addition or (in an actual engine) a longer period of fuel injection and, hence, more energy input. However, there is a limit to r_c. If the cut-off ratio is more than 10% of the stroke, smoking tends to occur in an actual engine, because there is not sufficient time for the combustion process to be completed before the exhaust valve opens. Thus, the products of combustion leave the cylinder before all the chemical energy has been extracted from the fuel. There is also a lower limit on r_c; there must be some heat or fuel consumption to

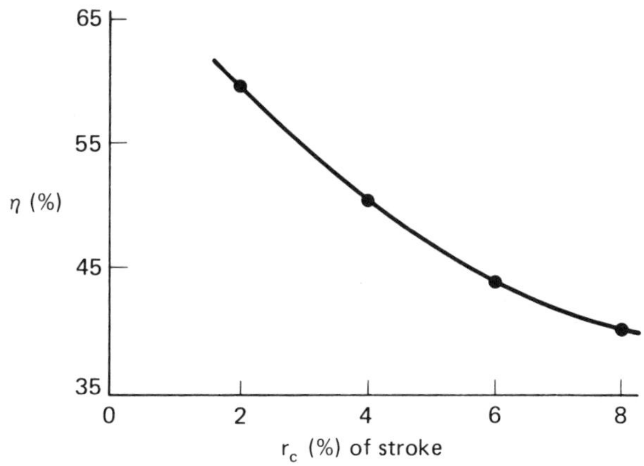

Figure 1-5. Cycle efficiency as a function of cut-off ratio at constant-compression ratio.

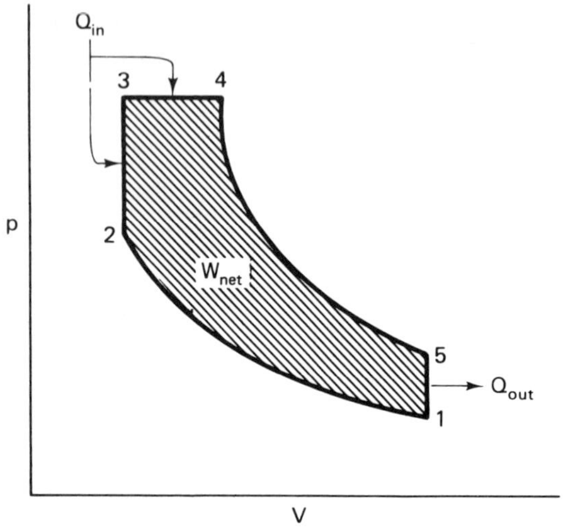

Figure 1-6. Air standard dual cycle.

overcome the friction of the various moving parts—the piston and cylinder and the friction associated with the bearings in the engine.

The *p-V* diagram of an actual engine does not follow the air standard *p-V* diagram. This is due in part to heat addition (injection) before and after top dead center. The air standard dual cycle corrects for the heat addition at other than constant pressure. In this cycle, illustrated in Figure 1-6, the heat addition occurs at both constant volume and constant pressure.

1.3 Actual Engine Cycles

The thermodynamic cycle, used for analyzing the air standard cycle, is not an actual engine cycle. It is impossible to recombust or reuse air, so a fresh air supply must continually be drawn into the engine and the products of combustion removed. Thus, in the actual engine cycle, unlike the thermodynamic cycle, the mass of air and the products of combustion undergo continual change.

Four-Stroke Cycle

Let us consider the four-stroke mechanical cycle found in diesel engines. Since the engine operates by compression ignition, at the top of each compression stroke a selected amount of fuel is injected into the cylinder and is ignited by the high air temperature. The combustion of the fuel continues as the piston gases expand on the power stroke, and the pressure remains essentially constant during the combustion process. Because there is only one power stroke for every two revolutions, it is necessary to know the operating cycle when calculating engine power. The four strokes may be classified as follows

1. Intake stroke. The intake valve is open, the exhaust valve is closed, and the piston moves down, bringing fresh air into the cylinder.
2. Compression stroke. Both intake and exhaust valves are closed, and the air is compressed by the upward movement of the piston.
3. Power stroke. Both intake and exhaust valves are closed and combustion occurs with a resultant increase in pressure, forcing the piston downward.
4. Exhaust stroke. The exhaust valve is open, the intake valve is closed, and the upward movement of the piston forces the products of combustion from the engine.

Since not all the products of combustion are removed from the cylinder on the exhaust stroke, there is a dilution of the incoming air charge by the remaining exhaust products. The greater the clearance volume, the greater the dilution.

The four-stroke cycle is a mechanical cycle, and there are two revolutions for every power cycle. One revolution is concerned with the intake and compression processes, the other with the power and exhaust processes.

Figure 1-7 illustrates a four-stroke cycle *p-V* diagram. The area marked W_I represents the work done by the engine. The W_{II} area represents the work consumed by the engine during the exhaust and intake stroke. The net work is the difference between the two areas. Area

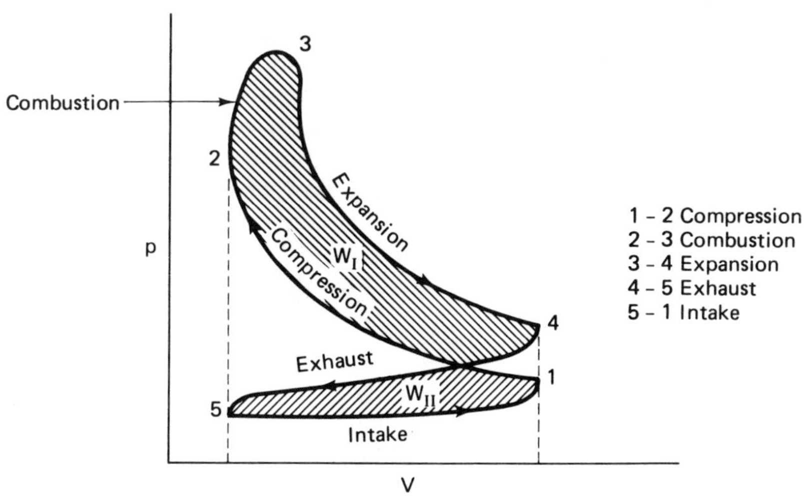

Figure 1-7. *p-V* diagram of four-stroke cycle.

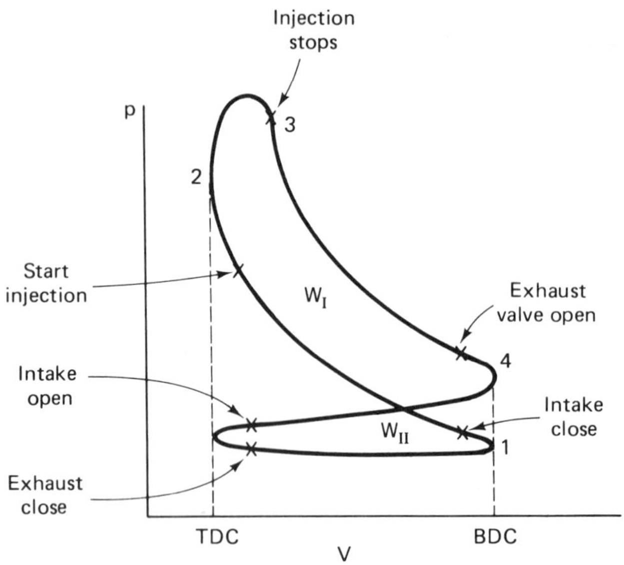

Figure 1-8. p-V diagram with valve openings and closings.

W_{II} has been exaggerated to show the effect of the exhaust and intake processes. There is some optimum value of exhaust valve opening which affects both areas. Point 4 represents the point where the exhaust valves open. Notice that if point 4 occurs earlier on the expansion stroke, less compressive work is required to push the exhaust gases out, as they flow under their own pressure. This reduces the value of W_{II}. However, less work is done by the gases pushing the piston, and so W_I is also reduced. There is no rule as to how soon or late the exhaust valves must open. For a given engine design finding the best values is a trial and error process. Of course, inlet and exhaust ducting directly affect the values of the W_{II}.

Figure 1-8 is also a *p-V* diagram for a four-stroke cycle engine, illustrating valve openings and closings and the limits of injection. Injection occurs before top dead center to allow time for the combustion process to begin. This time allowance is called *ignition lag*, the difference in time between injection and ignition. The fuel is ignited by the temperature of the air, so the fuel must first vaporize. When its temperature reaches ignition temperature, it ignites and combustion occurs.

Two-Stroke Cycle

The two-stroke cycle engine has a power stroke every revolution. Thus intake, compression, power, and exhaust processes must occur within each revolution. Figure 1-9 illustrates the *p-V* diagram for a two-stroke cycle engine that is not supercharged. To assist the exhaust process in the two-stroke cycle engine, the engines are equipped with scavenging blowers,

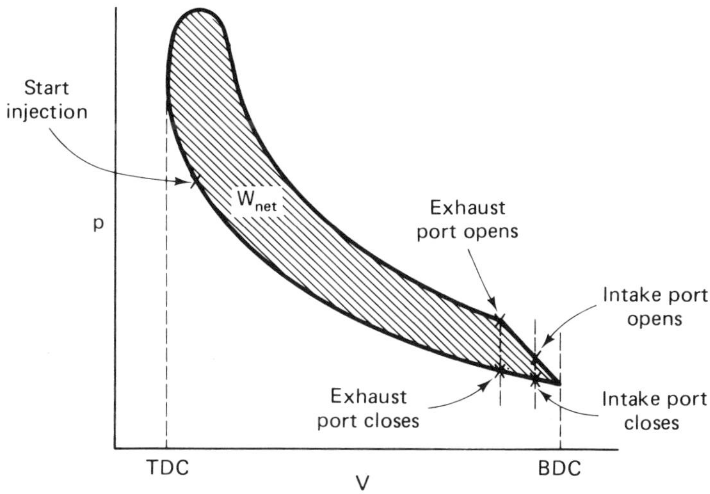

Figure 1-9. *p-V* diagram for two-stroke cycle (nonsupercharged).

which raise the inlet air pressure to 2–5 psi (13–35 kPa) above atmospheric pressure. Thus the intake air pushes the exhaust gas out. This is not the same as supercharging, which raises the inlet pressure much higher. The work required in operating the scavenging air pump or compressor is charged against the engine, so the net work of the two-stroke cycle engine is reduced by this amount. In addition to this loss of work, the combustion process often does not go as far toward completion as it does in the four-stroke cycle engine, so that the work of a two-stroke cycle engine is not twice, but only about $1\frac{1}{2}$ times, that of a four-stroke cycle engine.

1.4 Engine Indicator

How can a *p-V* diagram be obtained? On large, slow speed engines, a shut-off or indicator valve is provided on the cylinder head. An engine indicator can be attached, as illustrated in Figure 1-10. In this situation the gas pressure in the engine cylinder acts on the indicator piston. The motion of the piston is resisted by a calibrated spring. The spring is calibrated so that a unit motion of the stylus on the drum (1 in. or 1 cm) corresponds to a given value of pressure. The motion of the indicator piston, through linkages, moves the stylus arm. This gives the pressure axis of the *p-V* diagram. At the same time, a piece of cord connects the rotating drum to the reciprocating motion of the engine. The stroke of the engine is proportionally reduced, so the drum will rotate in an arc of less than 360°. This provides the volume axis of the *p-V* diagram.

The inertia associated with the engine indicator mechanisms creates

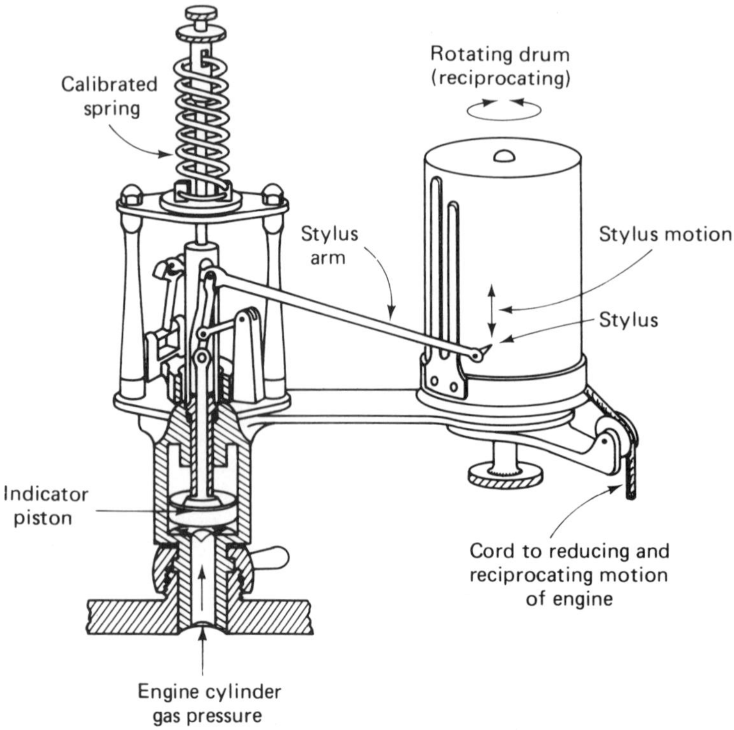

Figure 1-10. Engine indicator.

sufficient error in high speed engines so that the *p-V* diagrams are not representative of the engine behavior. The *p-V* diagram of a high speed engine may be found by using an oscilloscope with an electronic pressure pickup (pressure transducer). There are lightweight engine indicators for medium speed engines; however, pressure transducers may also be used. On certain engines there is no provision for translating the engine reciprocating motion to rotational motion in the *p-V* diagram. Special pressure gages may be used to determine peak cylinder pressure, which can also be used as a diagnostic tool.

The use of *p-V* diagrams is an important diagnostic tool; peak pressure and mean effective pressure readings are two direct benefits. To calculate the mean effective pressure, the area must be found, which at first glance seems a formidable task. However, a solution is available in the form of a planimeter, shown in Figure 1-11. The planimeter is a mechanical device which, when traced around the *p-V* diagram, will calculate the area. Then, by dividing by the length of the *p-V* diagram (representing the displacement volume), the mean effective pressure in inches or millimeters is found. Knowing the pressure scale, for instance, 1/4 in. = 50 psi, the mep may be determined.

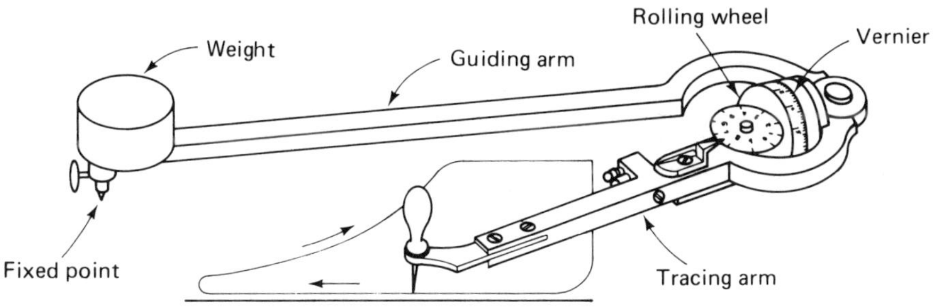

Figure 1-11. Polar planimeter.

The power may be calculated as noted in the earlier example by using mean effective pressure. We have just determined the *indicated* mean effective pressure (imep), calculated from the work the gas indicated it did on the piston. A *brake* mean effective pressure (bmep) may also be determined by

$$\text{bmep} = \frac{\text{Brake power}}{\text{Engine displacement}} \times \text{rpm} \qquad (1.11)$$

For example, consider an engine where the sum of all the cylinder displacements for an engine is 14 ft^3, the engine operates at 500 rpm, and produces 5000 hp. Determine the bmep.

$$\text{bmep} = \frac{(5000 \text{ hp})\ (33{,}000 \text{ ft-lb}_f/\text{min-hp})}{(14 \text{ ft}^3/\text{rev})\ (500 \text{ rev/min})} = 23{,}571 \text{ lb}_f/\text{ft}^2$$

If 23,571 lb$_f$/ft^2 is divided by 144, the number of in.^{2} per ft^2, the bmep = 163.7 lb$_f$/in.^{2} (psi). Comparing the bmep and the imep of an engine gives a clue as to the mechanical design of the engine. The closer the two values, the better the design as a greater portion of indicated work was converted to brake work. This technique is also useful when comparing different engines.

1.5 Volumetric Efficiency

During part of the diesel cycle the engine must act like a compressor and take air in and raise its pressure. The volume of air drawn in (during one stroke) is a function of the piston displacement. The term volumetric efficiency is used to describe how efficiently air is drawn into the engine. The ideal volumetric efficiency, η_V, is the ratio of the volume (mass) of air drawn in divided by the maximum possible amount of air that could be drawn in; i.e., the displacement volume (mass). Either mass or volume may be used in defining volumetric efficiency.

$$\eta_V = \frac{\text{Volume actual}}{\text{Displacement volume}} = \frac{\text{Mass actual}}{\text{Mass in displacement volume}} \tag{1.12}$$

If we let m_i be the mass actually drawn in, V_{PD} be the displacement volume, and ρ_i be the density of air drawn in, then

$$\eta_V = \frac{m_i}{\rho_i V_{PD}}$$

From the ideal gas law

$$\rho_i = \frac{p_i}{RT_i}$$

hence

$$\eta_V = \frac{m_i}{V_{PD}\,(p_i/RT_i)} \tag{1.13}$$

Several factors influence the density, and thus the mass, of air actually drawn into the cylinder. The pressure drop through the intake manifold and intake valve (port) reduces the pressure, and thus the density. Heating of the air from hot cylinder walls raises the temperature, which reduces the density. The effect of exhaust pressure and temperature is not directly included in this expression; however, the captured exhaust raises the average temperature, as well as diluting the pure air charge. The effect of altitude on volumetric efficiency may also be observed as the pressure and density decreases with increasing altitude. This means that less oxygen is available for the combustion process.

1.6 Cooling

Figure 1-12 represents a *p-V* diagram for a supercharged two-stroke cycle diesel engine. This engine is typical of large, low speed diesel engines. It is possible to develop the corresponding temperatures associated with compression and expansion processes in this engine. The temperature after combustion must be calculated by an energy analysis of the process. From the ideal gas law

$$p\,\frac{V}{T} = \text{Constant}$$

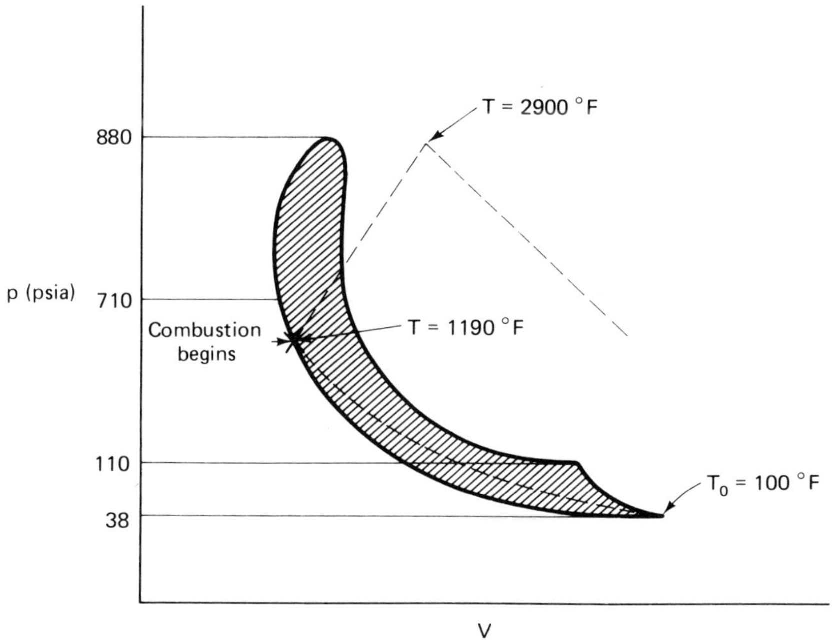

Figure 1-12. Pressure and temperatures found on two-stroke cycle p-V diagram.

or

$$T = T_o \frac{pV}{p_o V_o} \tag{1.14}$$

where the o state refers to the air at the beginning of compression. T_o would have a typical value of 100°F (40°C). The temperature distribution indicated by the dotted line in Figure 1-12 results. We see that very high temperatures occur within the cylinder. The following is a list of melting points of various metals

620°F	(327°C)	lead melts
1200°F	(649°C)	aluminum melts
2100°F	(1149°C)	cast iron melts
2700°F	(1482°C)	steel melts

A question logically follows: Why doesn't the engine melt? The reason is that the flame and the hot combustion gases do not have time to

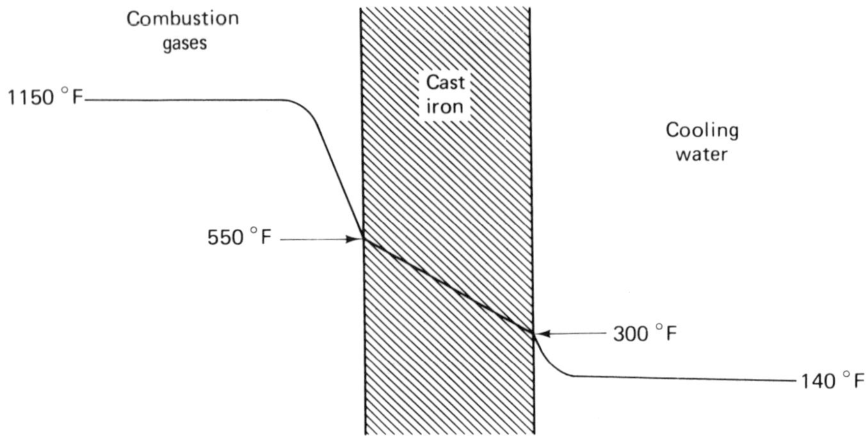

Figure 1-13. Typical temperature distribution across cylinder wall.

transfer enough energy (heat) to cylinder walls and pistons. For instance, in a large two-stroke cycle engine, about three cubic inches of fuel must be injected and burned every 1/400 of a second. Heat is transferred to the piston and cylinder, and these parts must be cooled or failure would eventually occur.

The average cylinder temperature is around 1150°F. Cast iron, the typical cylinder wall material, loses its strength at temperatures above 550°F. Figure 1-13 illustrates the temperature distribution through a cylinder wall. Fortunately, a large temperature drop occurs at the gas–wall interface. This is because the gas molecules form an insulating layer (boundary layer) next to the metal surface. The same effect is true on the cooling water side of the metal surface, but its effect is not so great because the density of the water is much greater. A similar problem exists at the top of the piston, and cooling must be provided there as well. Thus, cooling with the resultant insulating layers provides thermal protection to the metal surfaces of the cylinder.

1.7 Energy Balance

Let us look at the various energy flows entering and leaving an engine, as illustrated in Figure 1-14.

The first law of thermodynamics tells us that the energy entering the engine must be equal to the energy leaving. What, then, represents the energy of fluids (gas, liquid) entering or leaving the engine? This energy is called *enthalpy*. It is denoted by the letter h and has units of energy/unit mass (Btu/lb_m or kJ/kg). The mass flow rate is denoted by $\dot{m}$ (mass/time).

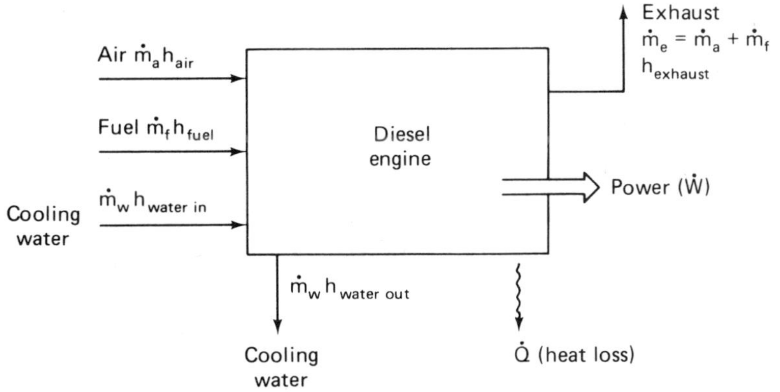

Figure 1-14. Energy flows in a diesel engine.

Thus, the energy balance is

$$\text{Energy in} = \text{Energy out}$$

$$\dot{m}_{\text{air}}h_{\text{air}} + \dot{m}_{\text{fuel}}h_{\text{fuel}} + \dot{m}_{\text{water}}h_{\text{water in}} = \dot{m}_{\text{exhaust}}h_{\text{exhaust}} + \dot{m}_{\text{water}}h_{\text{water out}} + \dot{W} + \dot{Q}$$

How is enthalpy measured? There are four measurable properties—temperature, pressure, volume and mass—so we should be able to determine enthalpy from one of these. The enthalpy of a gas and liquid may be written as

$$h = c_p T$$

where c_p is the specific heat at constant pressure and T is the absolute temperature in degrees Rankine or degrees kelvin. Table 1-2 lists the air, exhaust, and water specific heat values for several temperatures.

TABLE 1-2

	SPECIFIC HEAT	Btu/lb$_m$-°R (kJ/kg · K)	
		Temperature °R (°K)	
Substance	560 (311)	1060 (589)	1360 (755)
Air	0.24 (1.0047)	0.248 (1.038)	—
Exhaust	0.246 (1.030)	0.258 (1.080)	0.27 (1.130)
Water	1.0 (4.186)	—	—

The specific heat of a substance is a measure of its ability to absorb a certain amount of heat without increasing in temperature. For instance, it takes one Btu to raise one pound mass of water one degree in temperature, while it takes only 0.24 Btu to raise one pound of air one degree in temperature.

By using the indicated values of c_p in Table 1-2, we may calculate all the enthalpies—except one, the fuel enthalpy. The enthalpy value of the fuel is not quite so easily written, but when practical considerations are introduced, it too has a very amenable form

$$h = h_{RP} + (0.5T - 287)$$

The values of h_{RP} are listed in Table 1-3.

TABLE 1-3

Heating value (h_{RP})	ASTMD-975 No. 2-D	ASTMD-396 No. 4-D	No. 6
Higher Btu/lb_m @ 100°F	19,350	18,700	18,500
Lower Btu/lb_m @ 100°F	18,120	17,590	17,480

Thus, the enthalpy of No. 2 diesel fuel at 150°F (610°R) is $h_f = 19{,}350 + (0.5\,(610) - 287) = 19{,}368$ Btu/lb_m. The temperature effect, the term in parenthesis, is very small and may be neglected. Thus, $h_f \approx h_{RP}$, where the higher heating value is used.

The heating value is a measure of the heat liberated by the burning fuel and cooling the exhaust to 77°F. If the water vapor in the exhaust condenses, then the heat measured includes this effect, and is called the higher heating value. The lower heating value does not include the heat of the water that is condensed in the exhaust. Most small diesel engines use No. 2 diesel oil and many large, medium speed engines will use No. 4. The largest, slow speed engines may use No. 6 oil, after some fuel cleaning to eliminate harmful elements.

At this point, we should be able to calculate the heat balance for an engine. Our example engine receives 18,290 ft^3/min of air at 100°F and 14.7 psia and uses 2500 lb_m fuel/hour. The exhaust is at 870°F. The cooling water has a flow of 300 gpm, entering at 86°F and leaving at 164°F. The fuel is No. 2 diesel oil, with a specific gravity of 0.84, and the engine produces 7000 bhp. The engine has fourteen cylinders, each with a bore and stroke of 15.75 in. × 18.11 in., it runs at 520 rpm, and is a four-stroke cycle engine. Determine

1. The heat balance
2. Specific fuel consumption

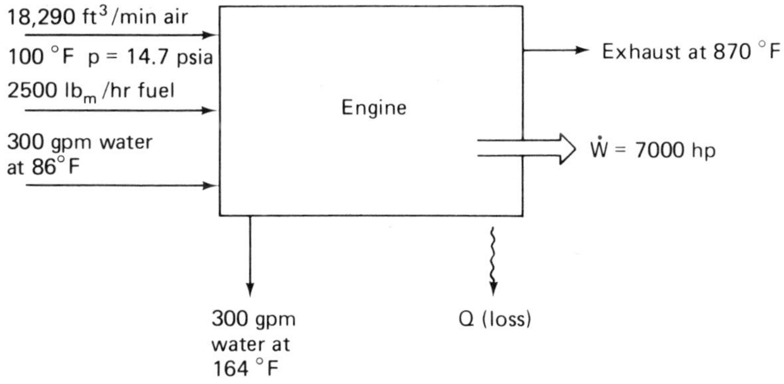

Figure 1-15. Sketch for example problem.

Figure 1-15 illustrates this engine with its energy flows. From the ideal gas law for air

$$\rho = p/RT = \frac{(14.7)\,(144)}{(53.34)\,(100 + 460)} = 0.0708\ \mathrm{lb_m/ft^3}$$

$$\dot{m}_{\mathrm{air}} = \rho_{\mathrm{air}} \dot{V}_{\mathrm{air}} = (0.0708)\,(18{,}290) = 1294.9\ \mathrm{lb_m/min}$$

$$\dot{m}_{\mathrm{air}} = 77{,}696\ \mathrm{lb_m\text{-}air/hour}$$

$$h_{\mathrm{air}} = c_p T = (0.24)\,(460 + 100) = 134.4\ \mathrm{Btu/lb_{m\,air}}$$

$$\dot{m}_{\mathrm{fuel}} = 2500\ \mathrm{lb_m\ fuel/hour}$$

$$\dot{m}_{\mathrm{exhaust}} = \dot{m}_{\mathrm{air}} + \dot{m}_{\mathrm{fuel}} = 80{,}196\ \mathrm{lb_m/hr}$$

$$h_{\mathrm{exhaust}} = c_p T = (0.27)\,(460 + 870) = 359.1\ \frac{\mathrm{Btu}}{\mathrm{lb_{m_{exhaust}}}}$$

$$\dot{m}_{\mathrm{water}} = (300)\ \mathrm{gal/min} \times 8.34\ \mathrm{lb_m/gal\ H_2O} = 2502\ \frac{\mathrm{lb_{m_{water}}}}{\mathrm{min}}$$

$$\dot{m}_{\mathrm{water}} = 150{,}120\ \mathrm{lb_m\ water/hour}$$

$$h_{\mathrm{water}} = c_p T = (1.0)\,(460 + 86) = 546\ \mathrm{Btu/lb_{m_{water}}}$$

$$h_{\mathrm{water}} = c_p T = (1.0)\,(460 + 164) = 624\ \mathrm{Btu/lb_{m_{water}}}$$

$$\dot{W} = 7000\ \mathrm{hp} \times 2545\ \mathrm{Btu/(hp\text{-}hr)} = 17{,}815{,}000\ \mathrm{Btu/hr}$$

According to the First Law

$$(77{,}696)\,(134.4) + (2500)\,(19{,}350) + (150{,}120)\,(546)$$

$$= (150{,}120)\,(624) + (80{,}196)\,(359.1) + 17{,}815{,}000 + \dot{Q}_{\mathrm{loss}}$$

$$\dot{Q}_{\mathrm{loss}} = 494{,}599\ \mathrm{Btu/hr}$$

Let us find that portion of the total energy that each group of energy terms accounts for. The energy input is ($\dot{m}_f h_f$) 48,375,000 Btu/hr. Energy loss to cooling water

$$\dot{Q}_{\text{cooling}} = \dot{m}_W (h_{\text{out}} - h_{\text{in}}) = (150{,}120)(624 - 546)$$
$$= 11{,}709{,}360 \text{ Btu/hr or } 24.2\%$$

Energy used as power

$$\dot{Q}_{\text{power}} = 17{,}815{,}000 \text{ or } 36.8\%$$

Energy loss in exhaust (above air inlet temperature)

$$\dot{m}_{\text{exh}} (h_{\text{exh out}}) - (\dot{m}_{\text{air}}\ 100^\circ\text{F}) = (80{,}196)(359.1) - (77{,}696)(134.4)$$
$$= 18{,}356{,}041 \text{ Btu/hr or } 37.9\%$$

Energy loss to engine room (plus unaccounted losses)

$$\dot{Q}_{\text{loss}} = 494{,}599 \text{ Btu/hr or } 1.1\%$$

Total

Power	36.8%
Cooling	24.2%
Exhaust	37.9%
Loss	1.1%
	100.0%

The *brake specific fuel consumption* (bsfc) is an indication of the engine performance and is the mass flow of fuel divided by the power produced. Thus,

$$\text{bsfc} = \frac{\dot{m}_f}{\dot{W}} = \frac{2500 \text{ lb}_\text{m}/\text{hr}}{7000 \text{ hp}} = 0.3571 \text{ lb}_\text{m}/\text{hp-hr}$$

The *brake specific air consumption* (bsac) is

$$\text{bsac} = \frac{\dot{m}_{\text{air}}}{\dot{W}} = \frac{77{,}696 \text{ lb}_\text{m}/\text{hr}}{7000 \text{ hp}} = 11.1 \frac{\text{lb}_\text{m}}{\text{hp-hr}}$$

The fuel/air ratio ($r_{f/a}$) is often used to indicate how well (lean) the engine is operating and is defined as

$$r_{f/a} = \frac{\dot{m}_f}{\dot{m}_a} = \frac{2500}{77{,}696} = 0.0322 \ \frac{\text{lb}_\text{m} \text{ fuel}}{\text{lb}_\text{m} \text{ air}}$$

Notice that bsac = bsfc/$r_{f/a}$.

Fuel is often given in gallons, a volume measurement, rather than pounds, a mass measurement. The specific gravity (sp. gr.) of the fuel, the ratio of its density to that of water, allows us to convert to the mass basic.

$$\text{Specific gravity} = \frac{\rho_{\text{fuel}}}{\rho_{\text{H}_2\text{O}}} = 0.84$$

If 8.34 lb_m water equals one gallon, then for the fuel in the example (0.84) (8.34) or 7.00 lb_m fuel equals one gallon. The fuel flow in gpm is

$$\frac{(2500 \text{ lb}_\text{m}/\text{hr})}{60 \text{ min/hr}} \ \frac{(1)}{7.00 \text{ lb}_\text{m}/\text{gal}} = 5.952 \text{ gal/min}$$

The previously described information will enable you to analyze and diagnose engine performance. The heat balance is a very useful tool, as energy must be conserved; if it does not appear in one form (power), it will appear in another (cooling water). Very often the heat balance will give an indication of trends if it is plotted with time. The exhaust gas may be used to produce steam in a waste heat boiler for hotel and auxiliary system functions. Its high temperature means that a great amount of useful energy is left over and may be used. The temperature of the exhaust leaving the stack must be sufficiently hot (300°F or hotter) to prevent condensation of the water vapor, which will cause corrosion of metal surfaces. The engines found today range in size from several horsepower to fifty thousand horsepower but the thermodynamic analysis and operating principles remain the same.

2

Fuels and Combustion

As Chapter 1 made clear, it is very important in operating a diesel engine to understand how and why energy is being transformed. An overriding concern today is fuel consumption, and the need to minimize fuel usage will grow with time. With this in mind, the importance of understanding fuels and the combustion process and the conversion of chemical to thermal energy is underlined.

2.1 Petroleum

Most diesel engines use a refined crude petroleum which has had some of its more volatile constitutents removed and does not include the heavy residual fuels. There are, of course, varying degrees of diesel oil, so let us start at the beginning, with the crude petroleum. The crude petroleum, or crude oil, is that found in a natural state beneath the earth's surface. It was formed over long periods of time by decayed organic matter, subjected to the high temperatures and pressures found within the earth's crust.

The composition of the crude varies from place to place as a function, in part, of the matter forming it. The primary components are carbon, hydrogen, and sulfur. There are traces of many other elements. Carbon normally accounts for 82–87%, hydrogen 10–14%, and sulfur (and trace elements) 1–6%. The heating value of the fuel is a function of the fuel's chemical composition; the greater the percentage of hydrogen, the greater the heating value.

To understand more completely how the oil gets from the ground to the diesel engine, let us examine some of the simpler methods of distilling the fuel. The crude oil contains many different hydrocarbons (C_xH_y), some of which are more valuable then others. By distilling the fuel, the more valuable (typically, the more volatile) hydrocarbons are evaporated and then collected. The process continues until a heavy oil residue remains. Elementary distillation processes rely on the difference in boiling points of the various hydrocarbons. When using such processes, 20% of the crude oil could be converted into gas or gasoline, 40% into distillate fuel and kerosene, and 40% into residual fuel. For economic reasons, it is preferable to have less residual and distillate fuel and more gas or gasoline. To this end, modifications of the straight-run distillation process have been devised. By introducing thermal cracking—breaking a hydrocarbon molecule into lighter and heavier hydrocarbon molecules by subjecting the crude to high temperatures and pressures—the distribution improves to 35% gas/gasoline, 35% distillate, and 30% residual. Finally, at least for our introductory view, catalytic cracking, or catalytic reforming, may be considered as well. In the catalytic cracking, a catalyst promotes the breakdown of residual hydrocarbons by high temperature and pressure. The catalyst does not enter into the reaction, but it will absorb certain molecules. In this case, the distribution rises to 48% gas/gasoline, 43% distillate, 7% residual, and 2% coke. Coke is carbon residue left by the cracking processes and is used in furnaces as a fuel.

2.2 Combustion Chemistry

The fuel must be burned, or oxidized, to transform its chemical energy to thermal energy. If we consider the fuel as containing only hydrogen and carbon, then the following illustrates the oxidation processes

$$C + O_2 \longrightarrow CO_2 + \text{heat}$$

$$H_2 + \tfrac{1}{2}O_2 \longrightarrow H_2O + \text{heat}$$

Both oxidation reactions are exothermic, resulting in the liberation of heat. There must, of course, be more to the reaction than just these three elements, or any fuel (wood, diesel oil, etc.) would spontaneously start burning. Indeed, there is. The first requirement is that sufficient oxygen must be present to start and continue the oxidation reaction; second, the fuel and air (oxygen) must be at a temperature which will allow ignition to occur. Temperatures of 400°C or 750°F are sufficient for the reaction to occur.

Having sufficient air for combustion is important for sustaining the reaction. Imagine a fuel droplet vaporizing, and the vaporized fuel mixing with the air and burning. The products of combustion (CO_2 and H_2O) now also surround the fuel droplet, tending to block the oxygen in the air from reacting with the carbon and hydrogen. Hence, the oxygen (air) must be supplied in greater quantities than are minimally required by the preceding oxidation equations. A number of hydrocarbon families are involved in the reaction as the fuel vaporizes, creating a thermal distillation process that leaves heavier chain molecules to the end.

The reaction process proceeds through several intermediary reactions before the final completely oxidized product is obtained. For instance, carbon will form carbon monoxide; then the carbon monoxide will form carbon dioxide. The oxygen will exist as O, and well as O_2. If at any time during the reaction combustion is stopped, a variety of substances will be found. Unburned hydrocarbons, carbon monoxide, carbon dioxide, water vapor, and several others will be found. Some of these substances are considered pollutants and have adverse environmental impacts.

2.3 Fuel Classification

There are many characteristics that are used to classify fuel. The following is a listing of some of the more important ones; there is no order of importance implied by the listing.

Specific Gravity: Defined as the ratio of the density of a substance to the density of water, both measured at 60°F.

API Gravity: An arbitrary type of specific gravity established by the American Petroleum Institute as

$$\text{Degrees API} = \frac{141.5}{\text{sp. gr. } 60/60^\circ\text{F}} - 131.5$$

When fuel is purchased on a volume basis, the specific and API gravities relate the volume to the weight of fuel. Furthermore, in general, the higher the API gravity of a fuel, the higher its cetane value. The cetane rating will be discussed later. In the Appendices, Table A-1 lists the specific gravity, degrees API, barrels per long ton, and pounds per U. S. gallon. If, for example, a fuel oil has a specific gravity of 0.9279, its degrees API is 21, and one gallon contains 7.727 pounds. It is not always possible to purchase fuel at 60°F. To correct for the variation of specific volume or density with temperature, Tables A-2a, A-2b, and A-2c may be used. Table A-2a is used for fuels with degrees API between 0–14.9; Table A-2b for degrees API between 15–34.8; and Table A-2c for degrees API between 35–50.9.

Water and Sediment: These impurities are obviously not useful and may be detrimental to engine operation. Water is the primary cause of corrosion in fuel storage tanks and equipment. Sediment may accumulate in storage tanks and clog filters.

Flash Point: The temperature to which the fuel must be heated to allow flammable vapors to form on its surface. These vapors can be ignited by an open flame held above the surface. The importance of the flash point is in storage safety, not performance in an engine. The flash point can be used by petroleum manufacturers and distributors as an indication of contamination; gasoline, for example, reduces the flash point. The minimum acceptable values are determined by classification societies and regulatory agencies. A flash point of 150°F would typically satisfy insurance regulatory specifications. As another example, the U. S. Coast Guard requires that the minimum flash point of light diesel fuel be 120°F.

Pour Point: The temperature at which oil ceases to flow. This temperature has relevance when the engine is used in cold regions, where the oil used should have a pour point 10–15°F below its operating temperature.

Ash: The ash content of oil is that material remaining after the fuel has been burned at a high temperature. Ash will accelerate engine wear through abrasion. The ash content of distillate fuels is quite low, essentially zero, and that of heavier residual fuels, typically less than one quarter of one percent.

Sulfur: This element is always present in crude oils and its constituents as a sulfur compound. Sulfur causes increased wear by reacting with other compounds to form deposits. In the exhaust system, any water will combine with the SO_2 to form an acid, thus creating corrosion problems. The same reaction can occur in the cylinder as water vapor exists as a product of combustion and a constituent of air.

Carbon Residue: This material is the solid carbon left after fuel is burned with a fixed amount of oxygen. This correlates with the carbon-forming characteristics of the fuel, which cause such problems as ring sticking and deposits within the combustion chamber. Slow speed engines, but not high speed engines, can use fuel with high carbon residue because of the comparatively longer combustion time.

Viscosity: Viscosity is a property of fluids and is a measure of resistance to flow. It is determined in a practical sense by measuring the time it takes a given quantity acted on by gravity to pass through an orifice.

Measuring Viscosity

Fluid viscosity is directly proportional to temperature. Thus, the temperature at which the viscosity is determined must be specified; for example, 70°F, 100°F, or 130°F, with the standard for marine use being 100°F. The viscosity is reported as having a certain value, say 500 seconds at 100°F (the time to collect 60 ml). That would be abbreviated 500 SSU, where SSU stands for Saybolt Seconds Universal. Values typically range between 32 and 1000; when oils have a viscosity greater than 1000 SSU, a large orifice is used and the viscosity is known as Saybolt Seconds Furol (SSF). Figure 2-1 illustrates a schematic of a Saybolt Universal Viscosimeter. Another system, the Redwood, measures a different amount and uses a different orifice size. There are two Redwood orifices, and thus two classifications, Redwood 1 and Redwood 2, but the principle remains the same.

Fortunately, relief from this range of systems appears to be in sight—the conversion to SI units. Of course, this involves another unit, called

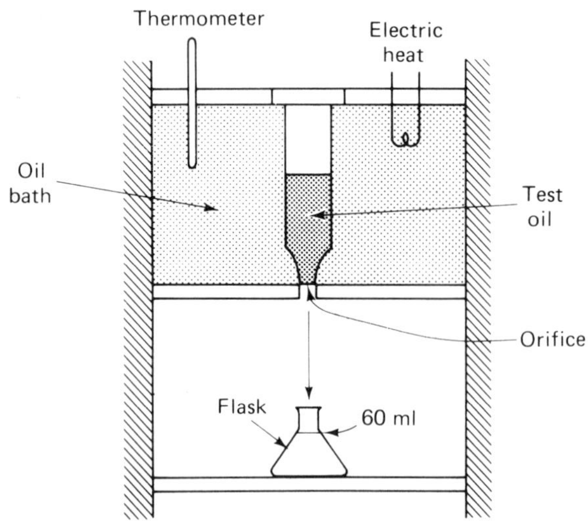

Figure 2-1. Schematic of a Saybolt Universal Viscosimeter.

TABLE 2-1 LIGHT MARINE FUEL OIL VISCOSITY

Redwood 1 @ 100°F	Centistokes (cSt) @ 50°C
200	30
300	40
400	60
600	80
800	100
1000	120
1200	150
1500	180
2000	240
2500	280
3000	320
3500	380
4000	420
4500	460

centistokes (cSt), for viscosity. In an attempt to minimize the potential confusion, Table 2-1 illustrates the old and new designations for light marine fuel oil.

Marine diesel oils have three different viscosities, all measured at 40°C. Table 2-2 equates the viscosity specifications of three marine diesel fuel oils.

TABLE 2-2 MARINE DIESEL OIL VISCOSITY

	Old	New
No. 1 D	45 Redwood 1, maximum @ 100°F	7.1 cSt, maximum @ 40°C
No. 2 D	40–55 Redwood 1 @ 100°F	5.3–10.3 cSt @ 40°C
No. 4 D	38–50 SSU @ 100°F	3.1–7.0 cSt @ 40°C

Heating Value: The measure of chemical energy which the fuel possesses. The heating value for a fuel sample may be calculated, if its specific gravity or API gravity is known, by

$$\text{Heating Value (Btu/lb}_m\text{)} = 18{,}250 + 40 \times \text{API gravity}$$

The heating value per pound increases as the API gravity increases, although the Btu's in a gallon decrease because of the decrease in mass per gallon. In terms of storage low API gravity fuels are desirable. Unfortunately, low API gravity fuels have high residues, lowering their combustion efficiency.

Fuel Oil Specification: Usually engine manufacturers specify a maximum Redwood 1 viscosity for fuel which may be used in the engine. 3500 is typical for slow and large medium speed engines. Remember, this is the time in seconds for 50 ml of oil to run through a standard orifice, nearly an hour per coffee cup. Thus, the 3500 Redwood 1 oil is very viscous and has a high carbon content and low API gravity.

Very viscous fuel cannot be injected into the cylinders so its viscosity must be decreased by preheating the fuel before it reaches the injectors. The injectors for a given engine will specify a certain Redwood 1 viscosity at the injector tip. Figures 2-2 and 2-3 are temperature, viscosity, and pressure charts that allow an engineer to adjust the fuel viscosity to meet the injector's requirements.

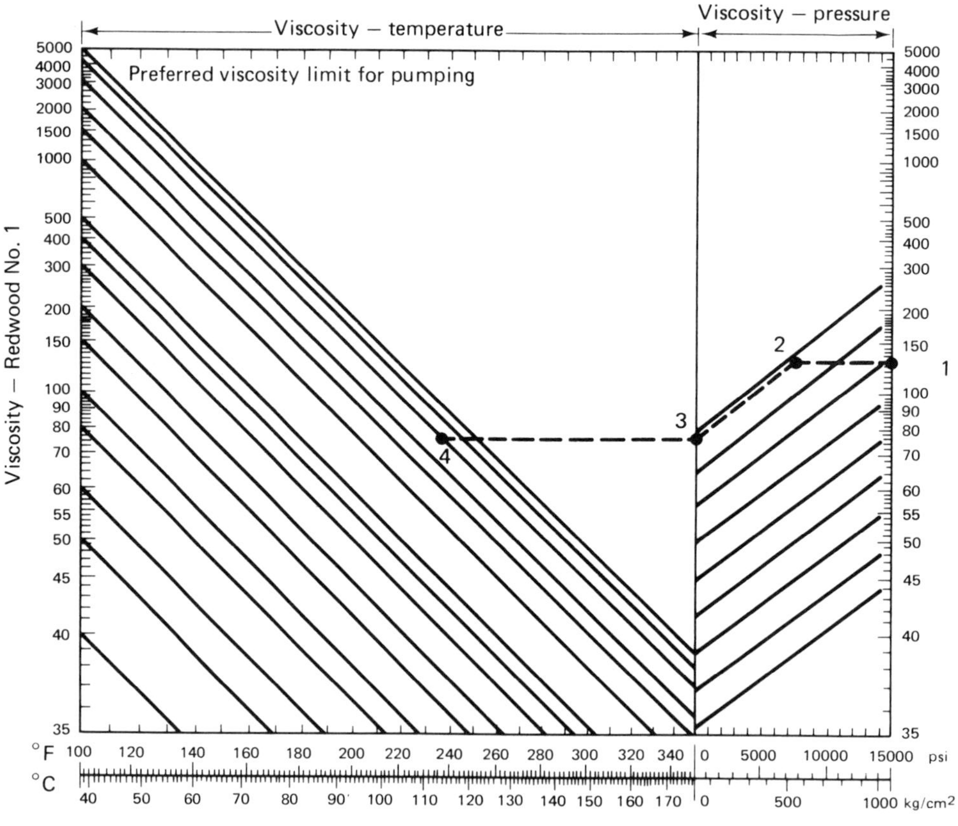

Figure 2-2. Viscosity, temperature, and pressure relationships for residual fuels. (Reproduced with permission from Mobil Oil Corporation.)

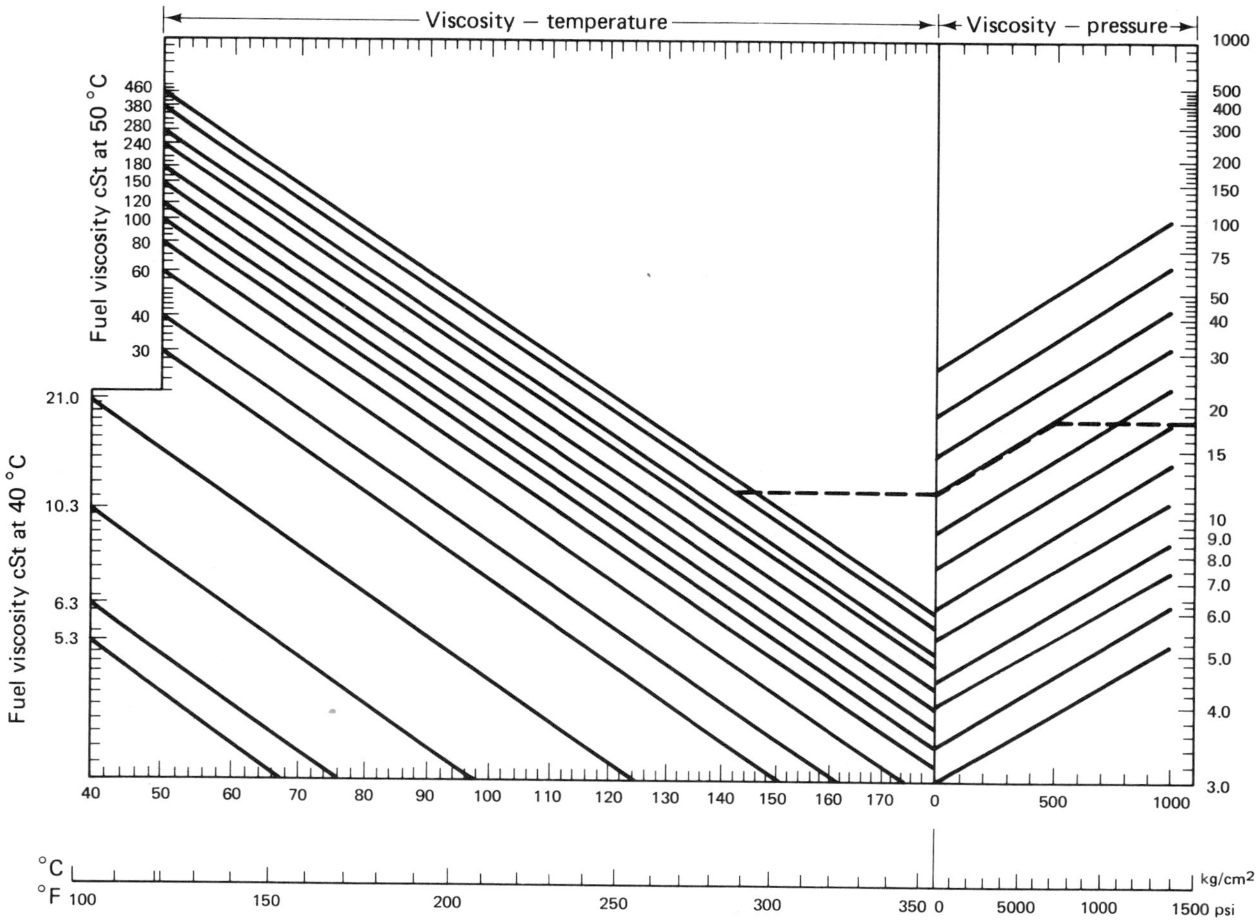

Figure 2.3. Viscosity, temperature, and pressure relationships for residual fuels. (Reproduced with permission from Mobil Oil Corporation.)

For example, a fuel oil received has a viscosity of 3000 Redwood 1 at 100°F, while the injector specification requires 130 Redwood 1. Refer to Figure 2-2 and follow the slanting line down 3000 Redwood until it intersects with the horizontal 130 Redwood line. Then the required temperature may be read on the x axis as 205°F. The fuel must be preheated to 205°F.

Viscosity is pressure dependent, and, since the fuel will be increased in pressure, the fuel temperature requirement must be modified to allow for this effect. In the previous example the injection pressure is known to be 7000 psi. Enter the chart at 130 Redwood on the right hand side and proceed horizontally to 7000 psi (line 1–2). Next, follow a path parallel to the diagonal lines to the temperature/viscosity diagram (line 2–3) and proceed horizontally to the intersection with the sloping 3000 Redwood 1 line (line 3–4). Read the temperature below; it is 236°F. Figure 2-3 provides the same information in centistokes. For example, if the fuel supplied had a viscosity of 380 cSt at 50°C and the burner required 18 cSt with an injection pressure of 500 kg/cm^2, the required temperature would be 142°C.

Table A-3 in the appendix lists the various conversions between viscosity indices.

2.4 Combustion Process

The air in a diesel engine normally is compressed to about 500 psi and 1000°F. Highly atomized fuel is injected into this hot air. The fuel begins vaporizing and there is a period of time between the injection and the beginning of the combustion process. This is known as ignition delay. Ignition delay occurs because the fuel must, first, vaporize and, second, have its temperature raised to the ignition point before combustion can occur. In a diesel engine, the combustion process starts throughout the entire combustion space, and the period between injection and ignition is called the delay period. The fuel is injected during the delay period, with a resulting fuel accumulation. At the beginning of combustion there is a very rapid combustion period with a rapid increase in pressure and temperature. The fuel injector is still injecting during this rapid combustion period. After the accumulated fuel is burned, the combustion period continues at a controlled rate; this is called the controlled burning period. Injection ends, but combustion continues until remaining fuel is burned. This period is called the after burning period. Figure 2-4 illustrates the pressure–time diagram for a typical diesel cycle.

2.5 Factors Affecting Combustion

Keeping the delay period as short as possible is of prime importance. A long delay period would mean a large fuel accumulation within the cylinder. Once the combustion process started, there would be very rapid in-

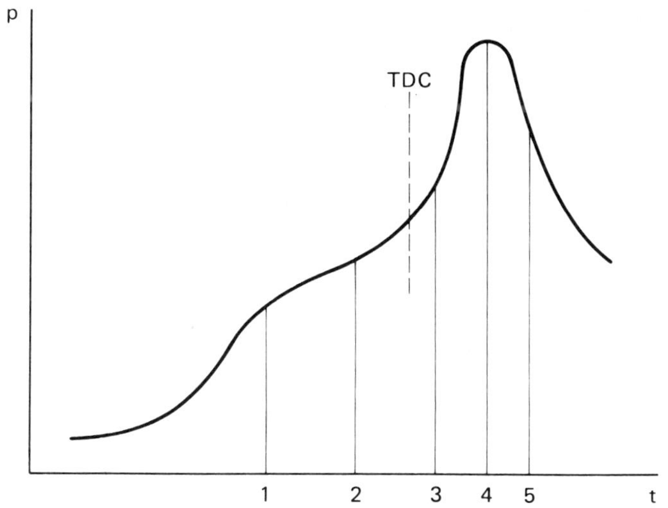

1 - Begin injection

2 - Begin combustion

t_{1-2} - Delay period

t_{2-3} - Rapid combustion period

t_{3-4} - Controlled combustion period

t_{4-5} - After burn period

4 - Injection ends

5 - Combustion ends

Figure 2-4. Pressure–time diagram.

creases in pressure and temperature, which would cause bearing and piston failures through mechanical and thermal stresses. The sharp pressure rise would contribute to very high impact loadings on the bearings and pistons and cause localized pitting and perhaps cracking.

There are several factors that affect the delay period

1. The cetane rating of the fuel. This is the ignition quality of the fuel, and it should be as high as possible. Normally, for a high speed engine, a cetane rating of 40 to 50 is required; the lower the cetane rating, the less ignitable the fuel. Conversely, the more easily the fuel is ignited, the shorter the delay period will be. The cetane rating takes into account such factors as the vaporization rate and the ignition temperature of the fuel.
2. Compression pressure. If the pressure is too low, the temperature will also be too low, and there will be a longer delay period while the vaporization process occurs. There must be at least partial vaporization before ignition may occur.

3. The timing of injection opening. If the injection opening occurs too early, then the fuel will enter the cylinder while the air temperature is too low. The vaporization time increases with a resultant large fuel accumulation.
4. The water jacket temperature. If the water jacket is too cool, it will lower the temperature of the air in the cylinder, increasing the delay period.

Intimate mixing of the air and the fuel is necessary for complete combustion. This requires a sufficient amount of air to burn the fuel droplets adequately. In a diesel engine, the air/fuel ratio varies on a mass basis between 20 and 1 and 100 to 1. The 20 to 1 is appropriate for higher loads; the 100 to 1 is for light loads.

3

Engine Construction

Engines may be classified in a variety of ways; a typical method is by engine construction. In this chapter we will start with the foundation of the engine, its bedplate, and conclude with the details of its superstructure, such as rocker arms.

3.1 Bedplates and Enbloc Construction

The two types of basic engine construction are bedplate construction and enbloc construction. There are two subdivisions of bedplate construction: box-type and marine-type. Since box-type engines may be found on board some vessels, both are explained. Figure 3-1 shows marine-type construction.

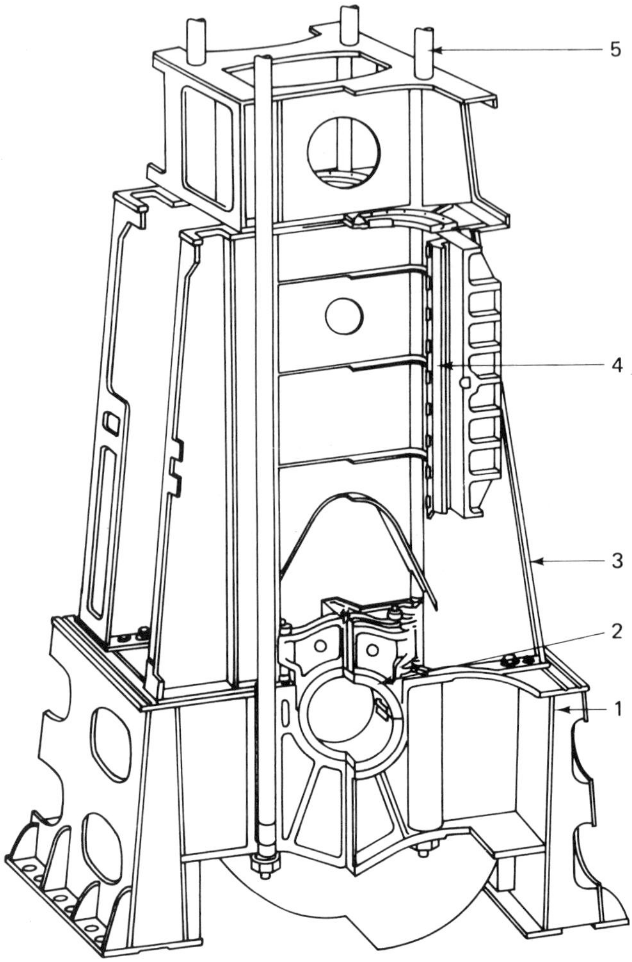

1 - Bedplate
2 - Upper main bearing and cap
3 - Engine frame
4 - Crosshead guide
5 - Through bolts

Figure 3-1. Bedplate and engine frame (marine type).

The bedplate is the foundation of the engine. Everything is built from it, so it must have sufficient strength to support all of the engine's fixed and moving parts. In Figure 3-1 the bedplate (1) is attached by its lower flanges to the ship's structure. The bedplate has a machined top surface and it supports the crankshaft directly. The crankshaft bearing caps (2) are illustrated in Figure 3-1, which is an example of block-type construction. In box-type construction, the oil is often contained within the lower part of the bedplate, an area called the crankcase. In the marine-type construction, the lubricating oil is stored in a separate area because the bedplate is very shallow.

The engine frame (3) is held in place by the through bolts, which also secure the engine block (not shown) to the frame. The crosshead guide will be examined in Section 3.12.

Figures 3-2(a) and (b) illustrate enbloc type construction. This is similar to an automotive engine. The crankshaft is hung from the block and below the crankshaft there is often a lightweight metal pan containing the lube oil.

The framework is bolted directly to the bedplate and supports the cylinder block and cylinder head. The head contains the valves and various other small pieces of moving equipment.

3.2 Liners

Cylinder wall liners used in diesel engines protect the engine block, which includes the bored cylinders, from piston wear. In automotive engines, the piston rings rub against the cylinder walls and gradually the rings wear the cylinder wall away. After time the compression decreases to such an extent that the cylinder must be rebored and fitted with new oversized pistons. This would not be practical for large diesel engines. The required lifetime of the engine is generally much longer than that of automotive engines and it is not desirable to have to rebore the entire engine periodically. The solution is to have liners that can wear out and be replaced. Once the wear limits have been exceeded, the liners may be removed and new units inserted, thus maintaining original equipment specifications.

There are three types of liners in marine use: dry, wet, and integral. Dry type liners use a counterbored cylinder with the liner installed in the bore of the cylinder. There is a tight seal between the liner and cylinder wall. Heat from the cylinder flows to the liner through the cylinder wall and into the water jacket. This type of liner is found only on small diesel engines.

Wet type liners also use counterbored cylinder blocks and there are gaskets on the top and on the bottom of the liner. The top gasket prevents water from leaking out of the water jacket. The bottom gaskets are O rings around the liner to prevent water from leaking into the crankcase. This type of liner has direct water contact with both the engine water jacket and the liner. Therefore, the heat flows from the cylinder to the liner to the water jacket. The wet liner is typically found on engines with bores greater than eight to ten inches.

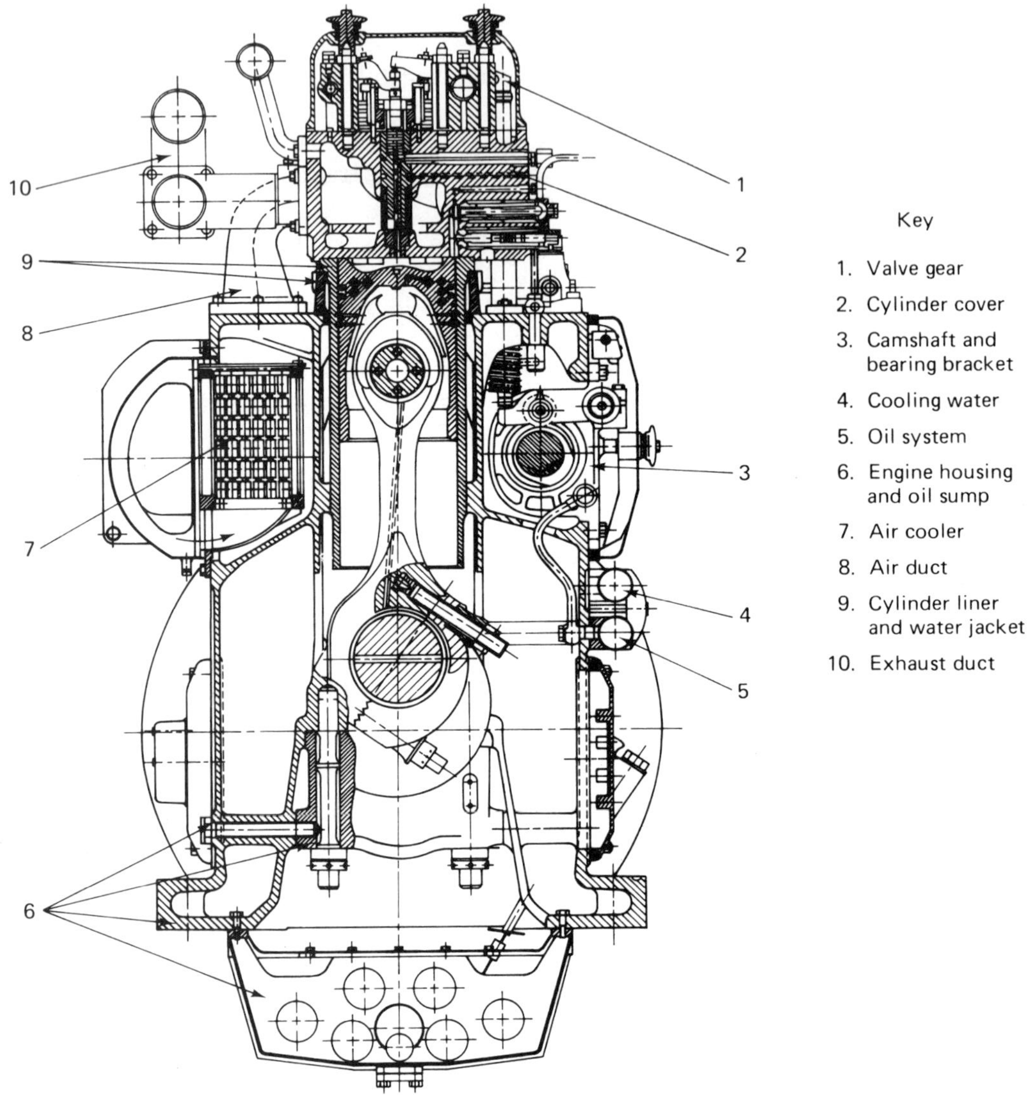

(a)

Figure 3-2. (a) Engine nomenclature (front view). (Reproduced with permission from Sulzer Bros., Inc.)

The integral liner also has water jacketing, but in this case the water jacketing is included in the liner. There is no water jacket in the engine block *per se*. The heat flow is from the cylinder to the liner to the liner water jacket. This type of liner is found in large General Motors engines and in Fairbanks Mores opposed piston engines.

1. Piston
2. Connecting rod
3. Fuel injection pump
4. Fuel system
5. Governor and governor drive
6. Camshaft and camshaft bearing bracket
7. Crankshaft
8. Flywheel
9. Main bearing
10. Oil system
11. Water system
12. Air ducts
13. Turbocharger

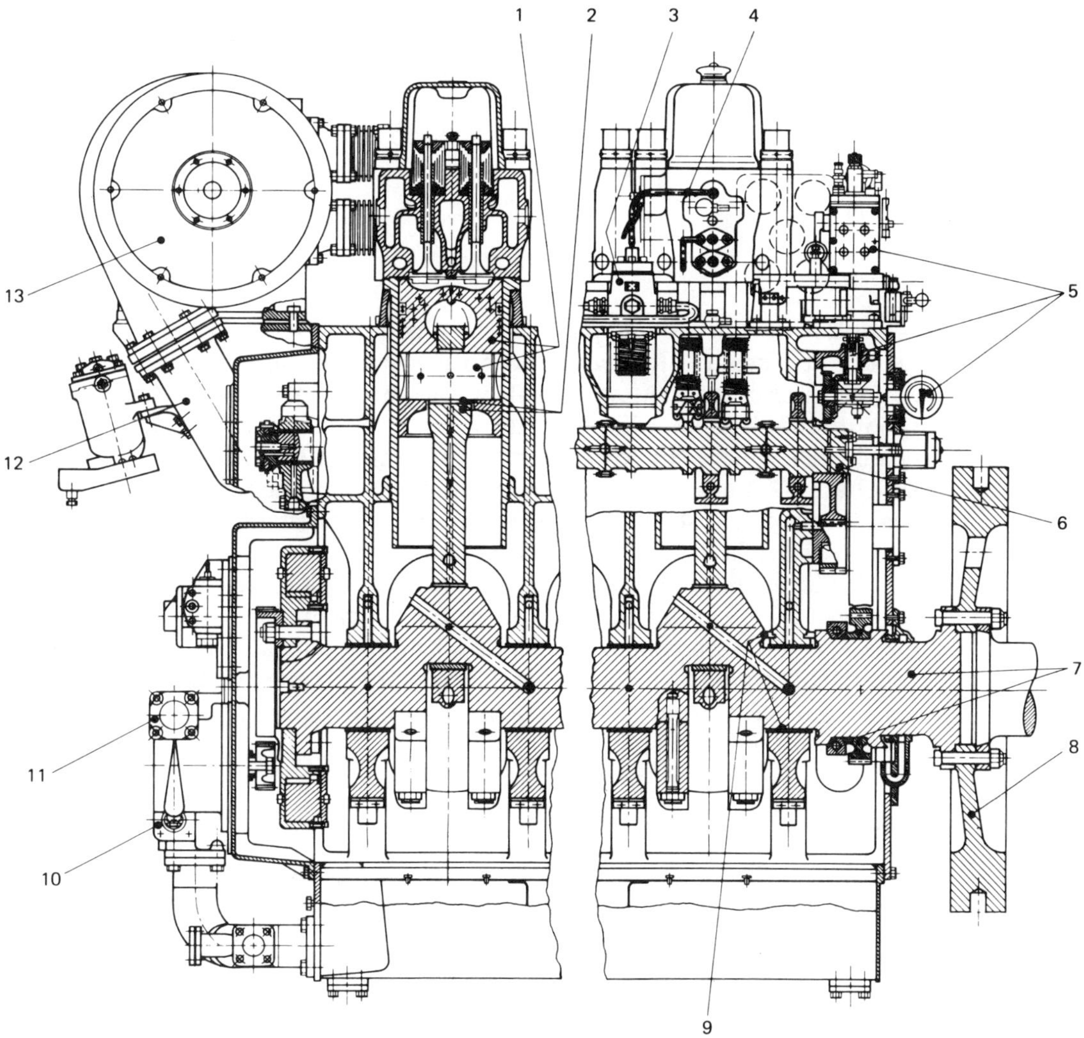

(b)

Figure 3-2. (b) Engine nomenclature (side view). (Reproduced with permission from Sulzer Bros., Inc.)

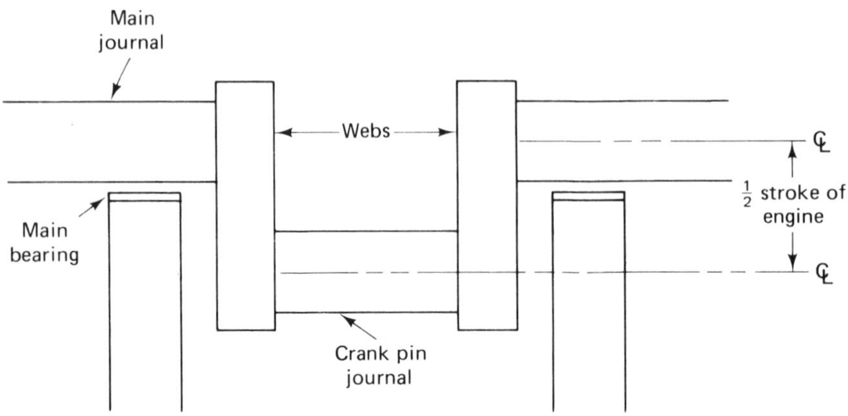

Figure 3-3. Crankshaft (side view).

3.3 Crankshafts

In considering the moving parts of the engine, we will start at the bottom, with the crankshaft. In Figure 3-3 notice the main journal (the horizontal part of the crankshaft), two crank webs, the main bearing, and the crank pin journal. The crank pin journal is also called the connecting rod journal. It is where the lower part of the connecting rod is fastened to the crankshaft. The crankshaft's construction may be of two types: one piece, found in most engines, or with a shrink fitting holding the webs to the main journals, found in Sulzer engines.

The term *crank throw* covers the two webs plus the crank pin journal. The length of the crank throw is measured from the centerline of the main journals to the centerline of the connecting rod journal. Thus, the vertical distance the web travels is equal to half the stroke of the engine; there must be sufficient space on the bottom of the engine, above the bedplate, to permit the webs to rotate.

3.4 Pistons

The trunk piston shown in Figure 3-4 is used in engines where the connecting rod goes directly from the piston to the crankshaft. The skirt of the piston must be able to withstand the side thrust forces developed by the angularity of the connecting rod movement. The upper part of the piston is slightly tapered to allow for expansion due to high combustion temperatures. The upper rings on the piston are the compression rings, which seal the combustion space during the compression process and power stroke to prevent leakage. The compression rings also serve to transfer heat from the piston to the cylinder liner. Slower speed engines require more compression rings than higher speed engines, because the

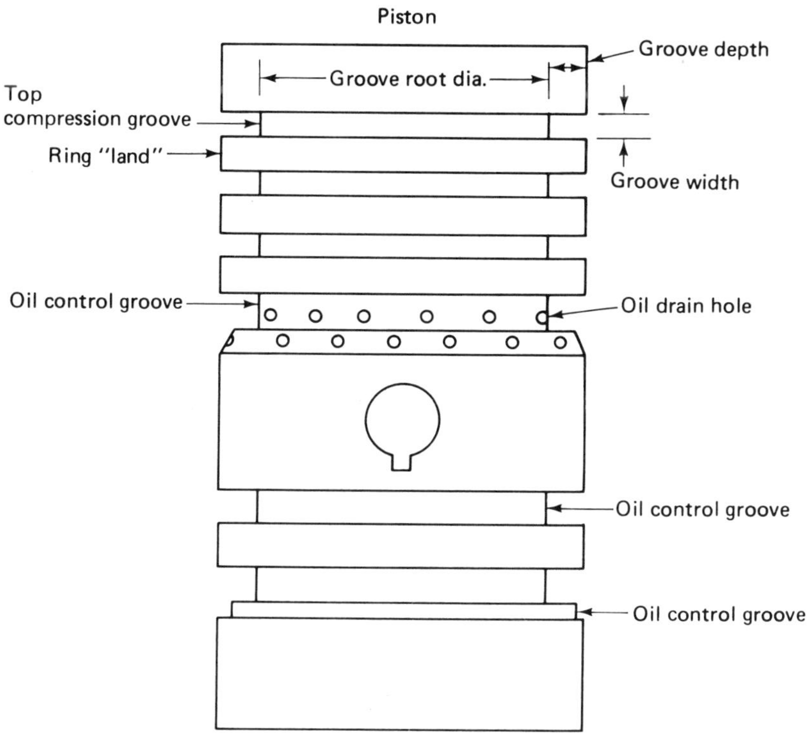

Figure 3-4. Trunk piston.

compression and power strokes take a longer time with the resultant greater air and combustion gas leakage. The oil rings are the lowest rings on the upper part of the piston. There also may be an oil ring on the lower part of the skirt of the piston. The oil rings' function is to remove excess oil from the cylinder wall and scrape it back into the crankcase. There are small drain holes through the piston behind the oil ring, which allow the oil to drain back to the crankcase. There would be no direct path for the oil flow back to the crankcase if these holes were not provided. Whenever pistons are pulled and the rings cleaned, it is important that drain holes are cleaned and opened.

3.5 Piston Rings

Figures 3-5(a) and (b) illustrate ring nomenclature and show the three clearances that must be considered when installing piston rings. Typically the first clearance to be measured is the ring end clearance or gap clearance. Prior to installing the ring on the piston, the ring is placed in the plane of least wear perpendicular to the cylinder axis. Thus, the ring would be in the liner as if it were on the piston. A feeler gage is placed

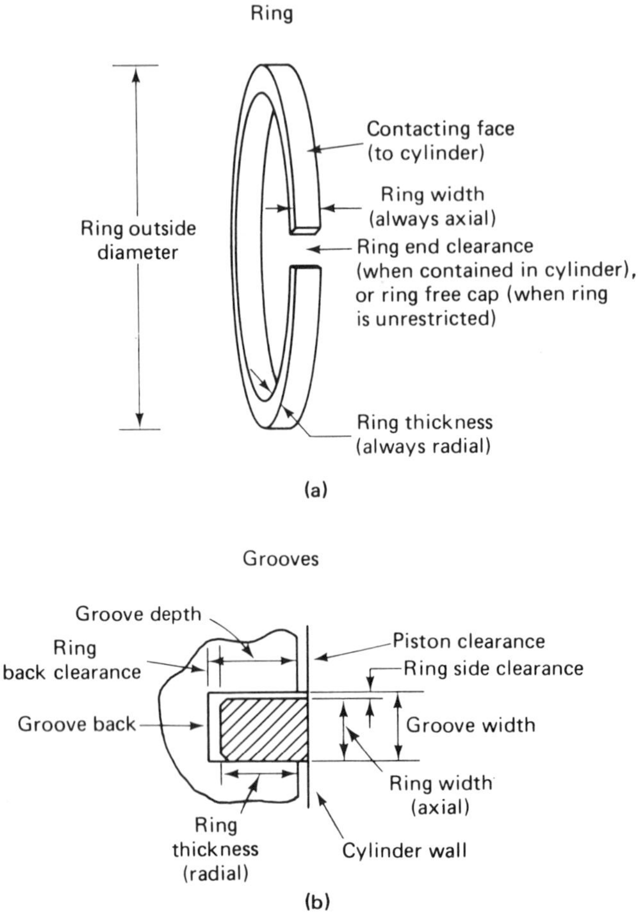

Figure 3-5. Piston ring and groove nomenclature.

between the ring ends and the distance is measured. The engine manufacturer specifies the minimum and maximum clearances. Should the clearance be less than the minimum, a possibility exists that the ends will butt due to thermal expansion of the ring on the hot engine. Butting ring ends would cause the ring to expand tightly against the liner, significantly increasing friction and the probability of piston seizure. Should the clearance be greater than the maximum specifications, combustion gases will blow-by the rings and exhaust gases will enter the crankcase. This will be accompanied by a loss of compression.

The second clearance is the ring side clearance, illustrated in Figure 3-5(b). This clearance is the distance from the top of the ring to the bottom of the ring land above the ring and is measured with a feeler gage. As before, the engine manufacturer specifies minimum and maximum clearances. If the clearance is less than the minimum, the combustion gases will not be able to get behind the ring and the ring will not be pushed against the liner for sealing purposes. Note that it is not only ring tension which maintains the seal, but gas pressure as well. In this case of less than minimum clearance the combustion gas will act on the contacting or outer ring face, compressing the ring and causing blow-by. If the clearance is too great, particularly on four-stroke cycle engines, excess ring groove wear will occur as a result of the ring snapping to the top of ring groove at the end of the exhaust stroke. This problem increases in severity as the engine speed rises, and is more common on high speed engines. In the two-stroke cycle engines the ring is always under pressure, so the problem is not significant.

The third clearance, which need not be measured, is called back clearance. With the ring in its groove, one side of the ring is pushed until it contacts the groove back. The ring thickness should be less than the groove depth, so the outer ring face is slightly inside the ring groove. If there is no back clearance, the rings will take the side thrust instead of the piston trunk, resulting in high friction and possible piston seizure.

Figures 3-6(a)–(j) illustrates a variety of compression and oil control rings. Compression rings are normally butt or angle cut (45° is common). Step cut rings are used to minimize blow-by at ring ends when the cylinder liner is worn. A bronze insert is used to allow a rapid wear-in and conformance of the ring to the cylinder. The quick seating ring accomplishes

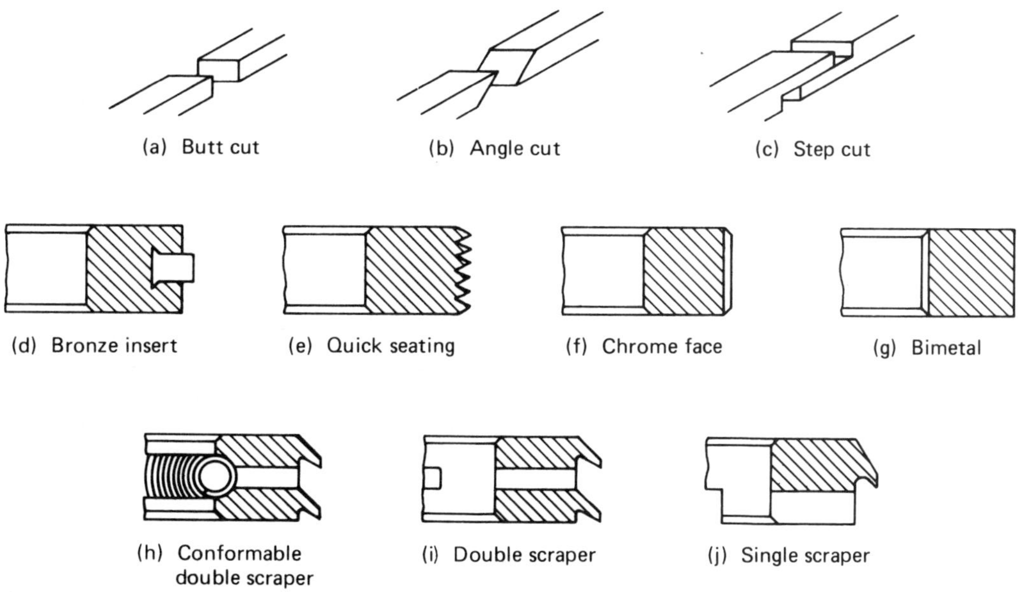

Figure 3-6. Ring types.

the same purpose by serrating the outer face, which will wear and conform quickly. The chrome-faced ring increases ring life as chrome wears slowly. Chrome rings cannot be used in a chromed cylinder liner as chrome is not porous and will not hold the oil film necessary for piston lubrication. Chrome rings are used on new cast iron liners where both the liner and rings are initially circular. The bimetal ring is cast iron with a strip of steel in the liner ring face to provide additional strength.

Oil control rings are frequently ventilated, which means that oil can pass through the ring to the ring back face and drain through the piston to the crankcase. Scraper edges remove the excess oil on the downward stroke and distribute the oil on the upward stroke. When the rings are installed, the scraper(s) must face down. If not, the ring will act as an oil pump, forcing lubricating oil into the combustion chamber. Rings may be single or double scraper types. In certain instances, where the ring does not have sufficient tension against the cylinder, a spring is placed behind the ring, providing adequate tension for proper oil control. This situation is illustrated in Figure 3-6(h).

3.6 Connecting Rods

Two types of connecting rods are found in the marine field: the automotive type and the marine type. They differ in the bearing connection on the lower portion. The automotive type is built in two pieces: the rod and the bearing cap. The bottom bearing cap is split to allow connection to the crankshaft. Marine-type connecting rods have three pieces: the rod, the upper bearing cap, and the lower bearing cap. The marine-type rod is connected to the piston pin as is the automotive type. The difference lies in how the bearing is connected to the crankshaft. In marine-type construction there is a bearing box that can be broken into two separate pieces. The connecting rod is attached to the top of the bearing box and joins the connecting rod with the crankshaft. Shims can be inserted under the foot of the rod to allow for bearing wear and to maintain the correct compression ratio in the engine. As the engine runs, there is gradual bearing wear, particularly on the upper half of the connecting rod bearing shell and on the lower half of the main bearing shells. This tends to lower the compression ratio, allowing a greater clearance volume at top dead center. To compensate for this, shims are installed so the volume at top dead center remains constant. Thus, the design compression ratio can be maintained.

3.7 Piston Pins

There are three types of piston pins. Full-floating piston pins are those in which the piston pin and the connecting rod can rotate independently. Semi-floating piston pins have the connecting rod rigidly fastened to the wrist pin; this allows no bearing material in the rod. Finally, there are fixed wrist pins where the connecting rod may move but the wrist pin is held rigidly. Fixed pins are used where oil cooling of the piston is required and oil passages go through the wrist pin to the piston.

3.8 Valves

The heads in diesel engines are either for four-stroke or two-stroke cycle engines. Four-stroke cycle engines have both inlet and exhaust valves in the head. Two-stroke cycle engines do not have intake valves but there may be exhaust valves. The purposes of the cylinder head are to seal the cylinder and hold the valves.

Figure 3-7 illustrates a typical valve and shows common terminology. The term *valve seat* refers to the contact area both on the valve and the

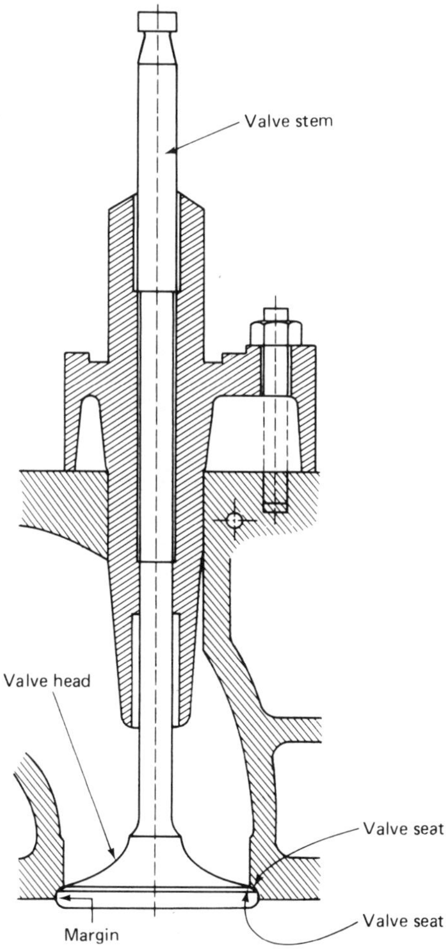

Figure 3-7. Valve construction.

cylinder head. Typically the valve seat is angled at 45° for diesel engines, although some valves use a narrower 30° angle from the horizontal. The 30° angle allows less restriction across the seat, and flow can start sooner and end later.

Figure 3-8 shows how the valve is held in the head. The valve guides are typically made of cast iron. The valves themselves are actuated by rocker arms which contribute some angularity to the up and down motion of the valve. The angular motion causes the valve stem to wear against the valve guides, which are replaceable. It is important that the clearance between the valve and the valve guide, called the *valve to valve guide clearance*, be within the manufacturer's specifications. If there is excessive wear, particularly on the exhaust valves, there may be carbon formation in the valve guides, as exhaust gases will collect in that area. Carbon forms and the valves become sticky and hang up. As the valves become sticky from carbon build up, they also tend to overheat, which further worsens the problem of valve deterioration. Excessive intake valve guide wear also causes high engine oil consumption because the lube oil from the rocker arm assembly is pulled through the guides on the piston's intake stroke.

The transfer of heat from the valves to the head occurs through the valve seat. The seats must be smooth, not only to prevent leakage, but also to allow for good heat transfer. Pits on the valve seats must be removed with lapping compound or, if the seats are very scored, they must be refaced. If an engine runs for a long time, normal wear will necessitate refacing the valve seats. Should this be done more than two or three times, the head valve seat diameter increases significantly, causing the valve to seat deeper in the head and restrict gas flow. Such a situation in an automobile engine would call for the heads being replaced. Marine diesel engines have valve seat inserts, shown in solid black in Figure 3-9,

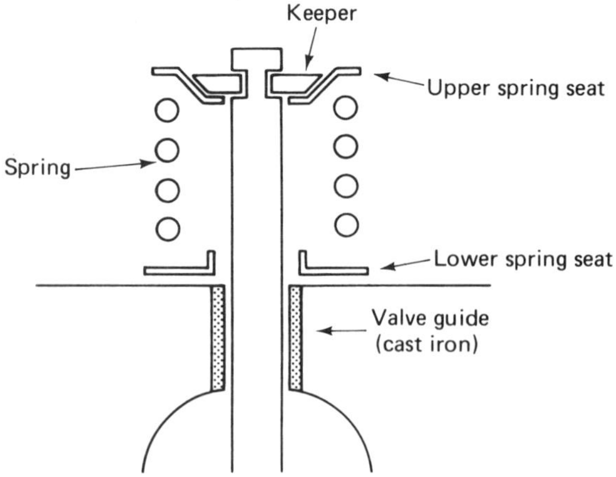

Figure 3-8. Valve guides.

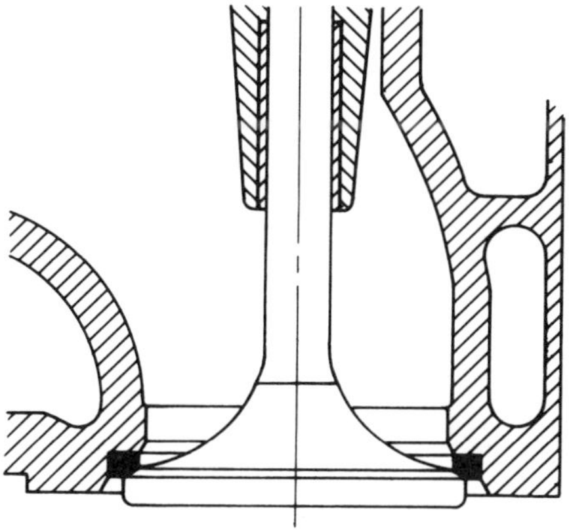

Figure 3.9. Valve seat insert (shown as solid black areas).

incorporated in the head itself. These are replaceable. As the inserts are ground away and exceed the dimensional limitations desired, they are replaced with new inserts. Color coded grinding stones are used for grinding valve seats. The color denotes the material the stone should be used on, e.g., cast iron, steel, or stellite seats.

There are limitations on grinding valves. One result of excessive grinding is the creation of too sharp an edge at the valve seat. This is created as the grinding causes the valve seat to incorporate material in the margin. Excessive temperature built up in this high heat stress area can cause the metal to fatigue and droop, a problem called tuliping, as the valve resembles a tulip!

3.9 Valve Cages

If a valve had to be repaired or replaced the entire head might have to be removed because the valves are installed solidly in the head. Heads can be removed, not easily perhaps, on automotive engines, but head removal (or pulling) is certainly very difficult on large diesel engines. *Valve cages* are used to overcome this difficulty. An example is illustrated in Figure 3-10. Each valve cage contains a valve assembly. A copper gasket is inserted between the combustion chamber and the cylinder and the head in valve cage. This copper gasket prevents both leakage from the combustion chamber and carbon build up. Leakage from the combustion chamber would be somewhat sooty, as it contains carbon particles. This carbon would gradually build up and essentially weld itself solidly between the head and the cage, making it very difficult to pull the cage. This is particularly true for exhaust valves. To help prevent welding or freezing up and to minimize uneven wear some exhaust valves are given a rotation; that is, the valve is rotated a slight amount with each revolution. Each valve cage for exhaust valves may have its own water jacket. The cylinder

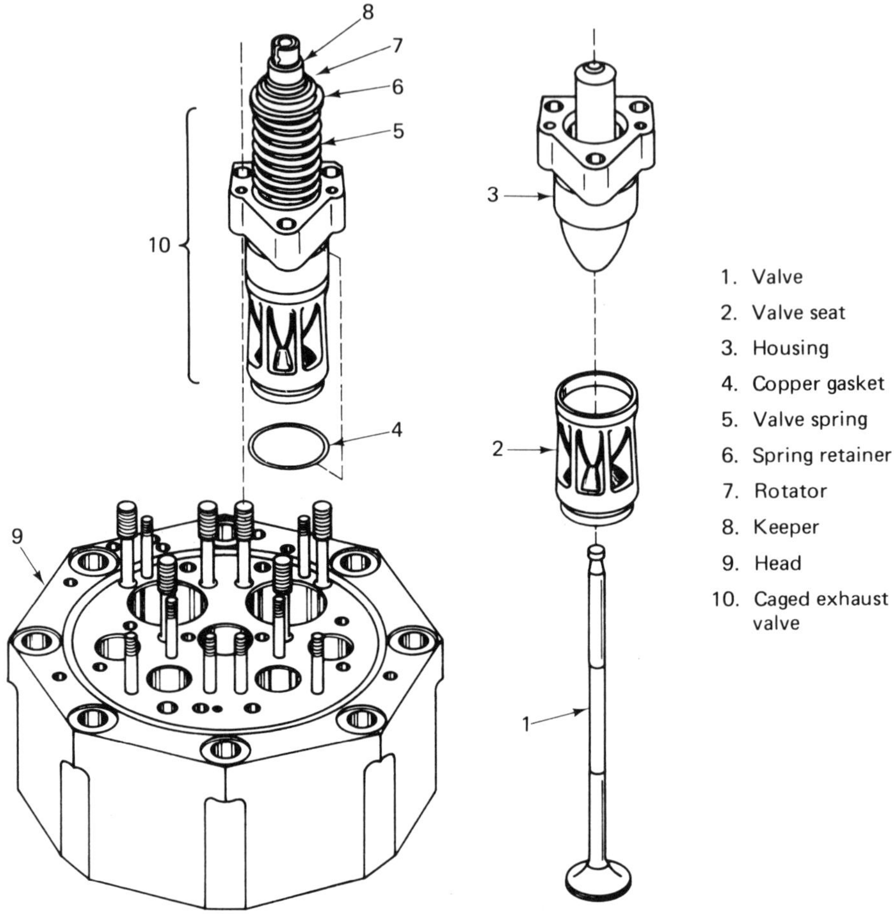

Figure 3-10. Caged exhaust valve. (Reproduced with permission from Colt Industries, Fairbanks Morse Engine Division.)

head has the cooling water for the intake valve, and heat transfer goes from the valve cage to the cylinder head to the water jacket.

3.10 Rocker Arms

Rocker arms are the devices that actuate the valves in the cylinder head. Figure 3-11 illustrates a typical rocker arm assembly. A cam, driven from the crankshaft, actuates the rocker arm through the cam follower and pushrod tappet. The rocker arm then pushes down on the valve stem. The rocker arm moves at an angle, primarily vertical, but with some horizontal thrust on the valve stem that causes wear on the valve guides. The

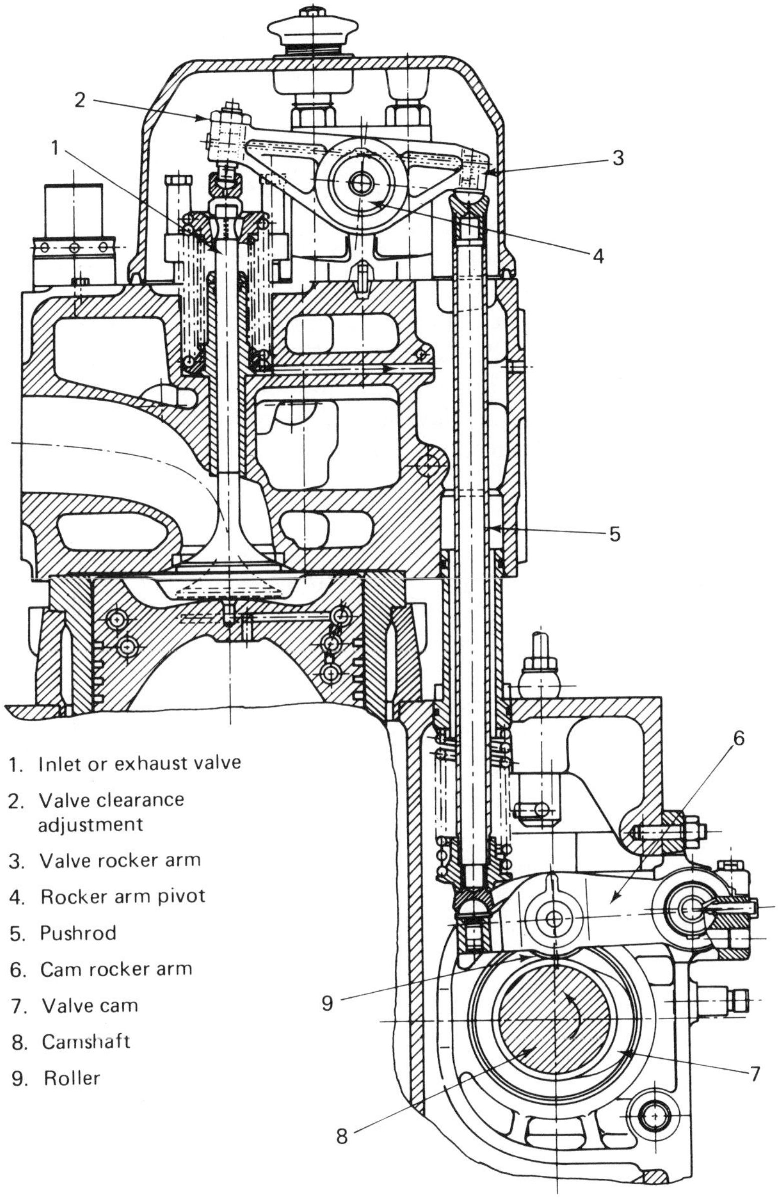

Figure 3-11. Valve actuation in the engine. (Reproduced with permission from Sulzer Bros., Inc.)

rocker arm assembly is attached to the head by a stanchion which is rigidly bolted to the head.

3.11 Cams

Figure 3-12 is an schematic illustration of a typical cam. Valves are timed while on the cam base circle. When the cam follower is riding on the base circle there should be clearance between the valve stem and the rocker arm. As the valve temperature increases, the valve will expand and, should there be no clearance between the valve and the rocker arm assembly, the valve would hang open. That would allow a constant flow through the valve, which cannot be allowed. If the clearance between the rocker arm and valve is too small, the valve will open early and close late. If the clearance is too large, the valve will open late and close too soon. In the next chapter we will see the effect of timing on engine performance.

3.12 Valve Clearance Adjustment

A feeler gage is used in conjunction with the tappet adjustment to set the valve clearance. To make an adjustment, a feeler gage is put between the valve and the rocker arm as the tappet is adjusted to the manufacturer's specifications for valve clearance. The cam, of course, should be on the base circle at this time. The term *ramp*, in Figure 3-12, denotes the distance required, moving on the cam flanks, to remove the valve clearance. The shorter the ramp, the less distance or time it takes for the clearance to be removed between the valve and the rocker arm.

Camshafts are driven by the crankshaft. There are three typical methods, all direct drive. The three drive types are gear drive, chain drive, and in some instances, vertical drive shaft. There can be no slippage, of course, as this could cause the engine to lose its timing.

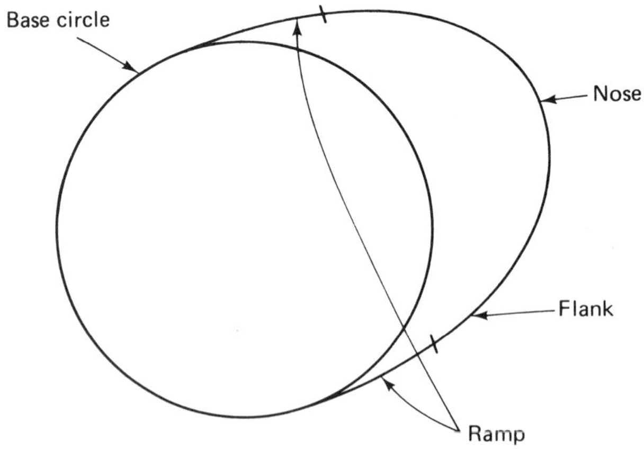

Figure 3-12. Schematic of cam.

3.13 Crosshead Engine

A second general type of marine engine, very often found in large, slow speed applications, is the crosshead engine. Figure 3-13 illustrates a typical crosshead engine assembly. The piston rod and, hence the piston moves only in the vertical direction; there is no side thrust, so the piston need not have an extended trunk area to absorb the thrust. There are, however, crosshead guides to take the side thrust, which results from the angularity of the connecting rod to crankshaft motion. The crosshead guides are supported by the framework of the engine. The pistons themselves may be of a relatively shorter length than those in trunk-type engine because all they contain are the compression rings for sealing and transferring heat. Typically there are no oil rings on crosshead engine pistons as the lubricating oil is partially burned, with the remainder being drained

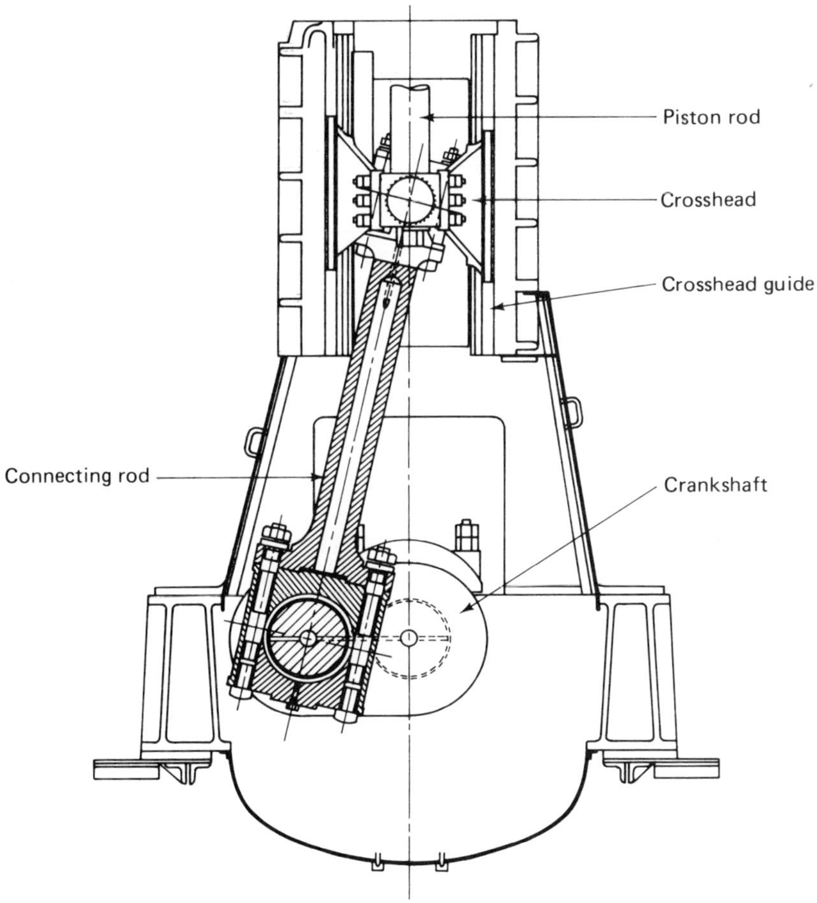

Figure 3-13. Crosshead engine.

away and not used again for lubrication. Crosshead engines require a very high alkaline oil because the fuel typically has a very high sulfur content. Large, slow speed engines are able to burn much lower grade fuel that has a high sulfur content. The high alkalinity of the lubricating oil neutralizes the acidity in the products of combustion, such as SO_2, that would cause corrosion of the pistons, piston rings, and cylinder surfaces.

4

Two-Stroke and Four-Stroke Engines

This chapter examines the mechanical aspects of two- and four-stroke cycle engines; how and why the engines operate as they do.

4.1 Polar Timing Diagrams

Figure 4-1 shows a polar timing diagram for a four-stroke cycle engine. The intake process occurs between points one and two, the compression process between two and three; the power stroke from three to four; and the exhaust stroke from four to five.

Figure 4-2 is a polar timing diagram for a two-stroke cycle engine; it is a schematic for a ported engine with intake and exhaust

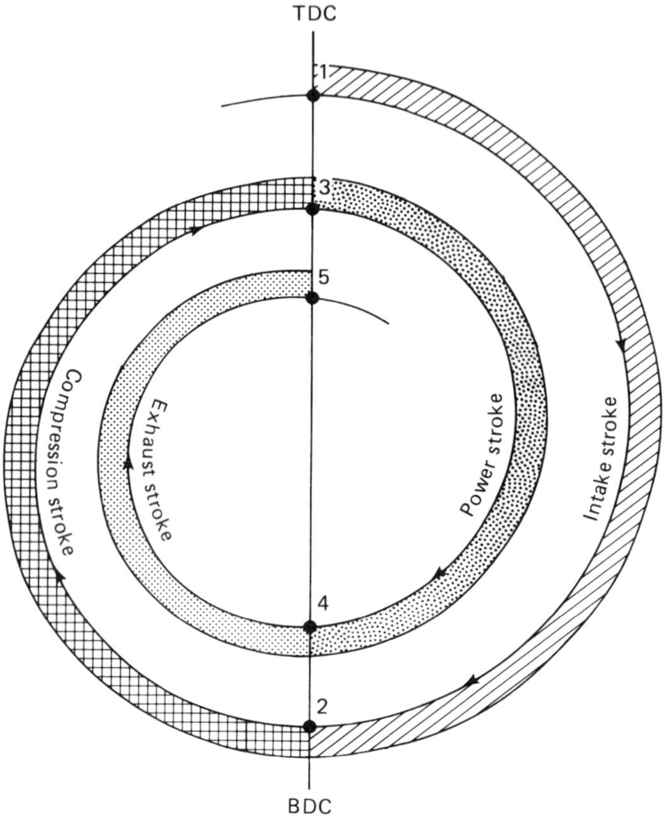

Figure 4-1. Polar timing diagram of a four-stroke cycle engine.

ports. Notice that the exhaust ports open first and close last, while the intake ports open last and close first. This allows a portion of the exhaust, under pressure, to leave the engine before the intake port opens and the intake charge enters the cylinder, scavenging the exhaust gases and recharging the cylinder with fresh air. On the upward stroke, the intake port closes first, before the exhaust port. The scavenging process is accomplished by either a scavenging air pump or a blower forcing intake air into the cylinder. The intake air pressure is usually two to five psi. Four-stroke cycle engines, on the other hand, are self aspirated. No scavenging air is required for those engines to operate.

The following list illustrates the valves that would typically be found in a four-stroke cycle engine head

1. Intake
2. Exhaust
3. Fuel injector

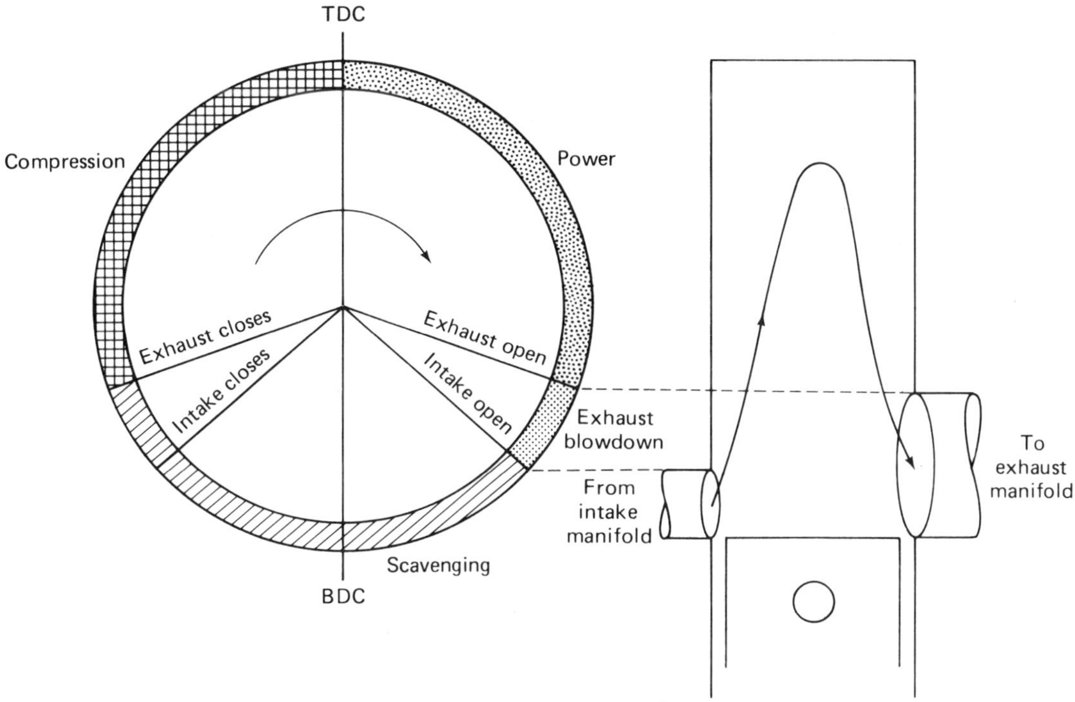

Figure 4-2. Polar timing diagram of a two-stroke cycle engine.

4. Relief valve
5. Indicator cock
6. Air starting valve

The valves on the two-stroke cycle engine are as follows

1. Never intake valves; always intake ports.
2. Possibly exhaust valves; very typically exhaust ports.
3. Fuel injector
4. Relief valve
5. Indicator cock
6. Air starting valve

4.2 Valve Timing

A typical polar timing diagram for a four-stroke cycle engine with exhaust and intake valve openings and closings is shown in Figure 4-3. Typically the intake valves open 5° to 20° before top dead center and close 20° to

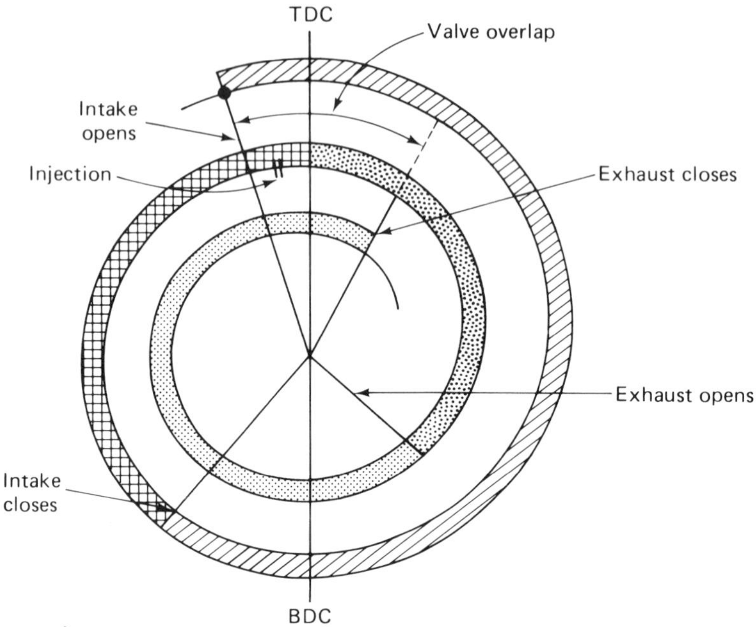

Figure 4-3. Polar timing diagram with valve overlap.

50° after bottom dead center. The exhaust valves open 50° to 20° before bottom dead center and close 5° to 20° after top dead center.

Why do the valve opening periods overlap? Consider the intake valve first. The cylinder is under reduced pressure compared to the intake manifold even after the piston has passed bottom dead center. This is due to the inertia of the air mass entering the cylinder. Also, as the piston is moving down, it reduces the cylinder pressure below that of the intake manifold. It is, however, desirable to allow the cylinder intake air pressure to build up to that of the intake manifold, increasing the air mass in the cylinder. There is very little vertical piston motion when the crank is crossing bottom dead center so the compression process is affected very slightly. The exhaust valve opens early to reduce the cylinder pressure and, hence, reduce the negative work required to push out the exhaust products. Any power that might be derived by delaying the exhaust valve openings is very small. Most of the vertical motion has been derived from the piston–crank assembly, and there is not much torque benefit from the piston's acting on the crankshaft at this point. Therefore, the exhaust valves open early, letting the pressure of the combustion gases in the cylinder flow into the exhaust manifold without pumping. This reduces the pumping requirement on the upward (exhaust) stroke of the piston. Notice that at top dead center there is valve overlap, where both the inlet and exhaust valves are open. This valve overlap period allows the inlet air to sweep the exhaust gases from the clearance volume of the engine and permits minimum dilution of the air charge in the engine. Table 4.1 illustrates the piston position as a percentage of stroke for increments of 10° from top dead center to bottom dead center. The table is arranged for several crank throw ratios, a topic covered in the next section. Notice that after 140° from top dead center, 90% of the vertical piston movement is accounted for.

TABLE 4.1 PISTON POSITIONS IN PERCENTAGE OF STROKE

Degrees from top center	Crank throw ratio L/R					
	3.75	4.00	4.25	4.50	4.75	5.00
10	1.0	1.0	0.9	0.9	0.9	0.9
20	3.8	3.7	3.7	3.7	3.6	3.6
30	8.4	8.3	8.2	8.1	8.0	8.0
40	14.5	14.3	14.1	14.0	13.9	13.8
50	21.8	21.5	21.3	21.1	21.0	20.8
60	30.0	29.7	29.4	29.2	29.0	28.8
70	38.8	38.5	38.1	37.9	37.6	37.4
80	47.8	47.4	47.0	46.7	46.5	46.2
90	56.7	56.3	56.0	55.6	55.3	55.1
100	65.2	64.8	64.4	64.1	63.9	63.6
110	73.0	72.7	72.4	72.1	71.8	71.5
120	80.0	79.7	79.4	79.2	79.0	78.8
130	86.1	85.8	85.6	85.4	85.2	85.1
140	91.1	90.9	90.7	90.6	90.5	90.4
150	95.0	94.9	94.8	94.7	94.6	94.6
160	97.8	97.7	97.7	97.6	97.6	97.6
170	99.4	99.4	99.4	99.4	99.4	99.4

Engine speed also affects valve timing. The range of valve openings is an indication of the speed effect. The exhaust valves of slow speed engines open at perhaps 20° before bottom dead center. The valves in higher speed engines open earlier, perhaps 50° before bottom dead center, to allow more time for the products of combustion to exit the cylinder and enter the exhaust manifold. The pistons on high speed engines move much faster, but the absolute time for certain processes (exhaust, intake) remains nearly constant, hence the need for a longer valve opening period. This is not the case in slow speed engines; the piston speed is much slower, and it is not necessary to have valve opening so early.

4.3 Crank Throw Ratio

Figures 4-4(a) and (b) illustrate the connecting rod crank throw ratio. Engines tend to have as small a crank throw ratio as possible. This allows the exhaust valve to open earlier and does not reduce engine performance. More positive work is available by having the lowest possible crank throw

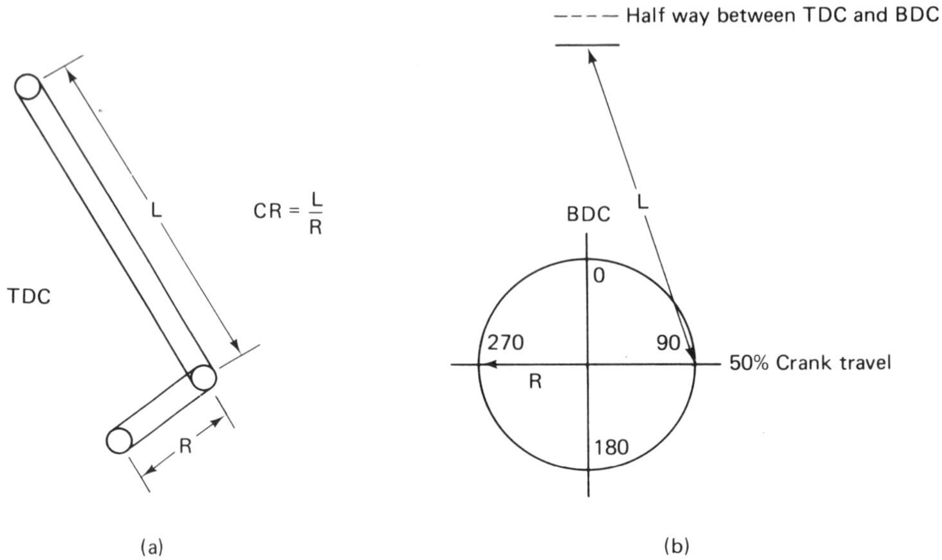

Figure 4-4. Crank throw ratio.

ratio, as more piston travel occurs during the first 90° of the cycle than during the second 90° period.

4.4 Firing Order

Figure 4-5 illustrates the firing order rotation of a six-cylinder engine. It is desirable that the same number of pistons fire during each revolution of the crankshaft and that the firing occur every 120°, allowing balanced forces in the engine. This means that there are always two pistons traveling in the same plane at the same time. For instance, consider pistons one and six. Number one may be at the end of the compression stroke and number six at the end of the exhaust stroke. They are 360° out of phase but travelling in the same direction. It is also desirable that pistons next to each other do not fire sequentially or simultaneously. If we consider these parameters—first, 120° apart in firing; second, that adjacent pistons do not fire together; third, that the number one cylinder usually fires first—then a logical firing sequence would be as illustrated in Figure 4-5 (one, five, three, six, two, four). Should we want to operate the engine in the opposite direction (in reverse), we determine the firing order by putting number one first and then reading backward from the previous firing order; in this case, it would be one, four, two, six, three, five. The correct angle between the firing of cylinders may be calculated easily by dividing 720° by the number of cylinders. This rule of thumb is valid for all except for two-cylinder engines.

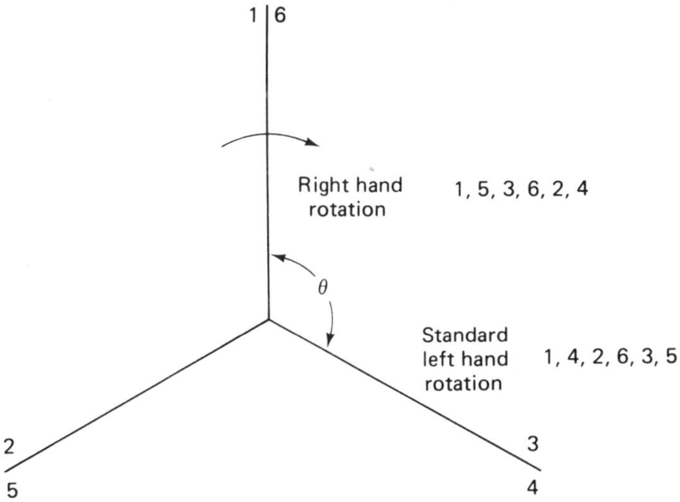

Figure 4-5. Standard six-cylinder engine firing rotations.

4.5 Cam Circle Diagram

Figure 4-6 shows the cam circle diagram for one cylinder of a six-cylinder engine. The cam must open three cylinders during one revolution and the next three cylinders during its next revolution. The camshaft itself turns at half the crankshaft speed. The overlap of the exhaust and intake valves is typically symmetrical on four-stroke cycle engines. In other words, both valves remain open for about 10°, or some other specified angle.

4.6 Overlap Method of Cam Timing

The installation of a camshaft is possible even though the timing marks are unknown, because of the equal overlap. This kind of situation can occur if a camshaft is broken at sea; a new camshaft arrives but there are no timing marks on it. This method presumes that if you time the number one cylinder correctly, all the other will be correctly timed. The method is called the overlap method of timing and it proceeds as follows: Insert the camshaft into the engine and put it in the overlap position. Insert the cam followers and push rods in the number one cylinder. Mount a dial indicator on the intake push rod and jack the engine over slightly backwards in the reverse direction until the push rod no longer moves. The dial indicator will not give a reading. This means that the cam follower is on the base circle of the cam. Place a reference mark on the camshaft gear case housing as shown in Figure 4-7. Rotate the engine in the direction of forward rotation; the cam follower will start to rise. As soon as the push rod moves up, stop the engine and mark the gear next

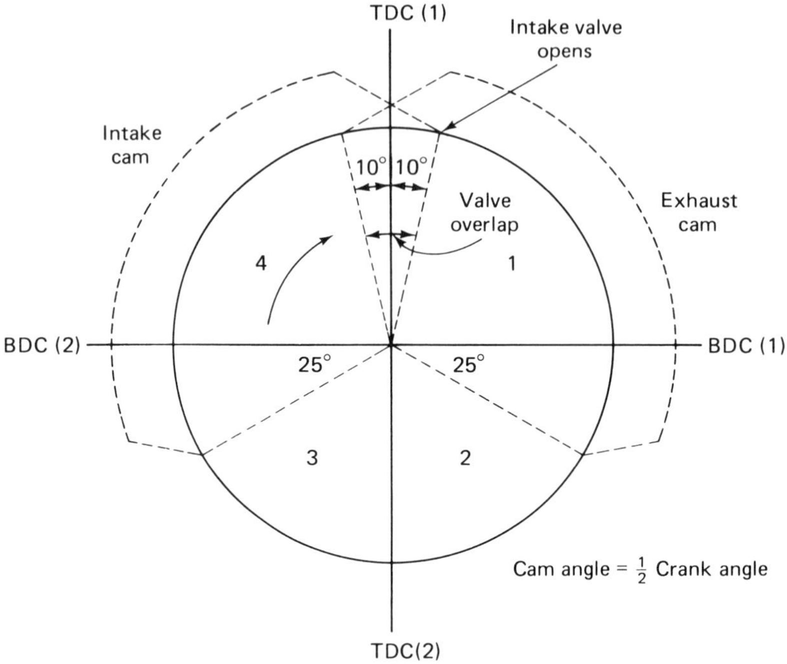

Figure 4-6. Cam circle diagram.

to the reference mark. This reference mark will be the beginning of intake or intake valve opening. Now put the dial indicator on the exhaust push rod and jack the engine over in the direction of rotation until there is no movement in the dial indicator. This again means that the cam follower is on the cam base circle. Jack the engine in the reverse direction slowly until the dial indicator moves again. Place a mark on the gear next to the reference mark. This indicates the exhaust valve closing. Halfway

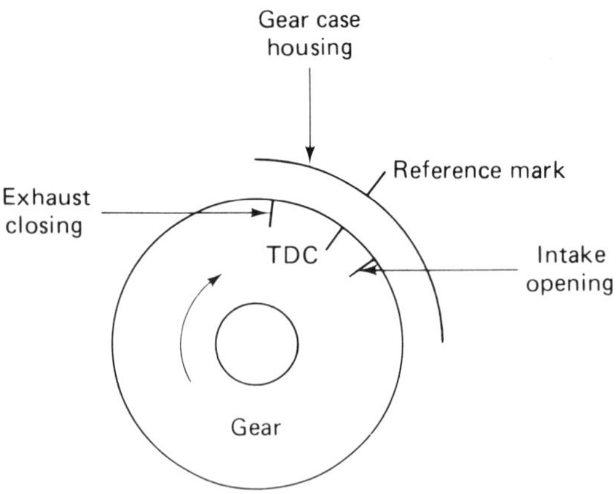

Figure 4-7. Markings used in the overlap method of timing.

between the exhaust valve closing and the intake valve opening is top dead center. Refer to the polar timing diagram to verify this. Mark top dead center on the gear halfway between the intake valve opening and the exhaust valve closing. Unmesh the cam and crankshafts and spot the number one piston on top dead center. Spotting the cylinder on top dead center again might have to be accomplished with the dial indicator, but the engine will have to be barred over. Reengage the camshaft gear so that the top dead center mark lines up with the reference mark. Now we have the number one cylinder on top dead center with the camshaft indicating top dead center and the engine in time. This is valid for four-stroke cycle engines.

4.7 Starting Air Valves

Large engines are air started. Starting air is normally admitted between TDC and 5° before TDC. There are some variations depending on the engine manufacturer. The starting air must be admitted on the power stroke of the engine to the minimum number of cylinders required to start an engine from any position. Qualitatively, there must be enough angularity on the power stroke to start engine rotation. If there are fewer than five cylinders in a four-stroke cycle engine, the pistons must be spotted in such a way as to create the momentum necessary to start the engine. If the pistons are not intended to be spotted, there must be a minimum of five cylinders in the engine. In a more quantitative sense, there needs to be an overlap of about 20° for air starting between the various cylinders. Consider that for a six-cylinder engine the minimum amount of the power stroke would be six divided into 720° or 120° plus 20°, which gives 140° per cylinder for air starting. Multipling 140° by five gives a cycle of 700°. Thus, if just five cylinders of the six possible were used for air starting, the engine would not be able to start if it stopped in the 20° not covered by air starting. Six cylinders for air starting are needed in this instance, thus providing 140° X 6, or 840°, of air starting. That engine could start from any cylinder position.

4.8 Two-Stroke Cycle Timing

Two-stroke cycle engines are not self aspirated and need a blower for scavenging. Figure 4-8 shows various methods of scavenging. Figure 4-8(a) shows direct or crossflow scavenging; the intake and exhaust ports are opposite one another. The inlet air stream is deflected by the piston shape and scavenges the exhaust air. Loop scavenging is illustrated in Figure 4-8(b). In this case both the inlet and exhaust ports are on the same side; the piston shape directs the air flow in a loop from the intake to the exhaust port.

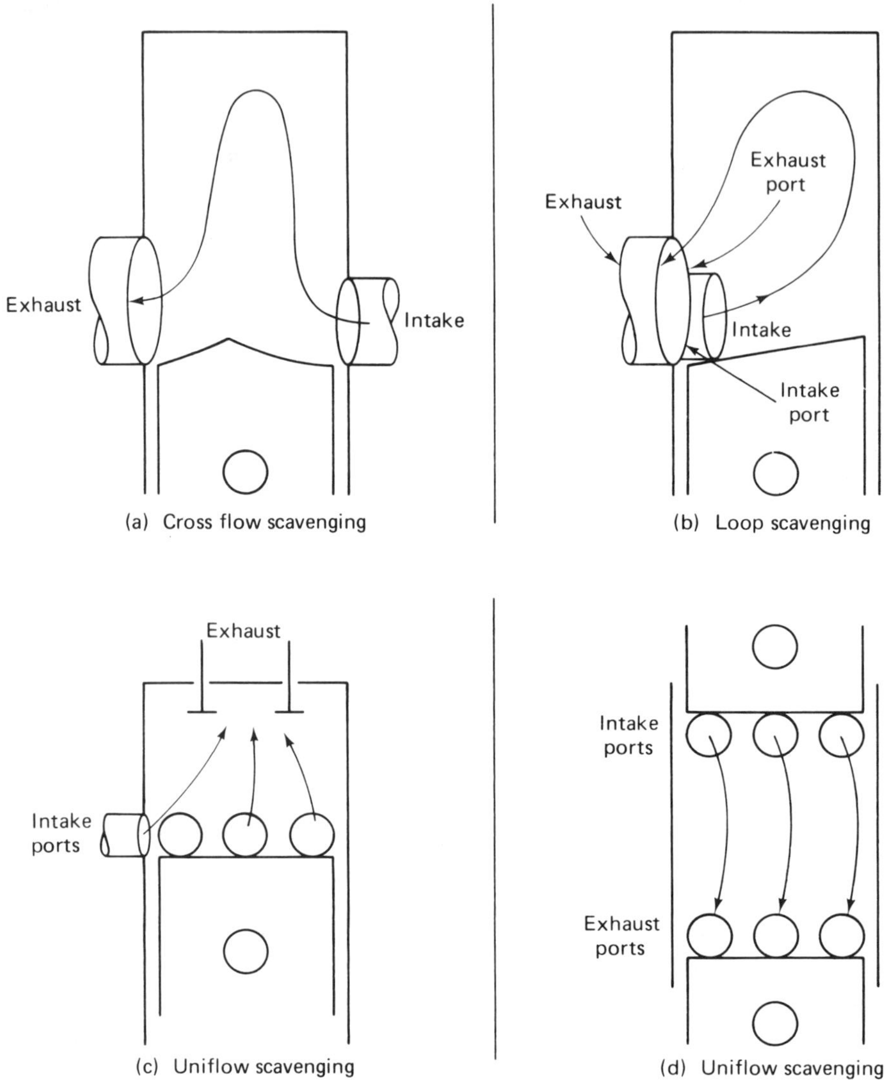

Figure 4-8. Types of scavenging.

Figure 4-8(c) illustrates uniflow scavenging. In this case there are intake ports and exhaust valves. Engines typical of this type of construction would be large General Motors and Burmeister and Wain engines. The intake ports are located around the periphery of the engine and the exhaust valves are in the head. Air flow is in a uniform direction vertically out of the engine.

Figure 4-8(d), illustrating an opposed piston engine, is another method of uniflow scavenging. In this case the intake ports are shown on the top and the exhaust ports on the bottom, although they may be reversed. In either case, the air flow is from intake port to exhaust port. There is a flow in one direction only, hence the name uniflow.

Figure 4-9 is a polar timing diagram for a two-stroke cycle engine. As before, limits are given on port openings and closings. The faster the engine rpm, the earlier the ports must open, to allow time for scavenging and recharging the cylinder with fresh air.

Air starting systems for two-stroke cycle engines require a minimum of four cylinders in the engine. This is because the cycle is 360° rather than the 720° of the four-stroke cycle. If only three cylinders were used for air starting, and the exhaust ports were open at 120°, the engine would not have enough momentum to start. The air starting valves on these engines are typically cam operated, as may be the fuel pump and exhaust valves.

The cam turns one revolution for every two revolutions of the crank in four-stroke cycle engines, while in two-stroke cycle engines they have equal rpm and, hence, the cam and crankshaft have the same angle of rotation. The angle required for air starting of two-stroke cycle engines is calculated as follows: Divide 360° by the number of cylinders and add 20°; this yields 110° for the duration of air starting for the four-cylinder engine. For a six-cylinder engine the figures would be 60° plus 20°, or 80° duration for the air starting cycle.

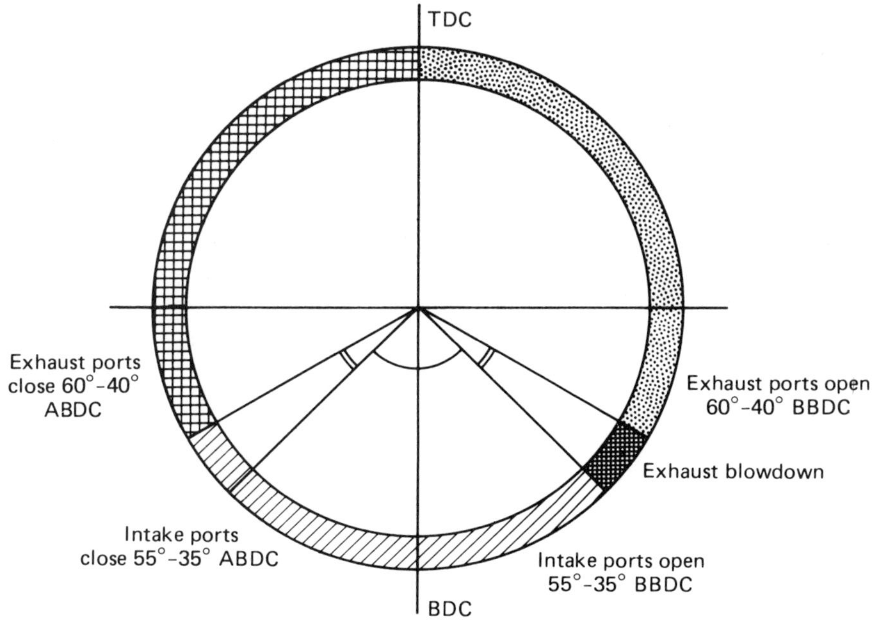

Figure 4-9. Two-stroke cycle polar timing diagram with port openings and closings.

4.9 Speed Classification

Diesel engines may also be classified by speed. High speed diesel engines have an rpm greater than 1000, medium speed engines are between 300 and 1000 rpm, and slow speed engines run at less than 300 rpm. There is also a speed classification based on piston speed. The piston speed is measured in feet per minute.

$$\text{Piston speed} = \frac{2 \times \text{Stroke (in.)} \times \text{rpm}}{12} \ \frac{\text{ft}}{\text{min}}$$

A third method of speed classification is the speed factor. It is defined as

$$\text{Speed factor} = \frac{\text{Piston speed} \times \text{rpm}}{100{,}000}$$

The following classifications apply to the various speed factors:

3 or below — slow speed
3 to 9 — medium speed
9 to 27 — high speed
27 and above — super high speed

5

Combustion Chambers and Related Systems

5.1 Introduction

The combustion process is the key to engine operation. Research has found that turbulence in the combustion process greatly aids the combustion efficiency. Therefore, combustion chambers are designed to promote turbulence.

Mixing of the fuel and air is very important to achieving proper combustion. In two-stroke cycle engines with exhaust valves, the intake ports are cut tangentially to swirl the inlet air. As the compression process occurs, the swirl increases. The tangential energy forces the air into a circular rotation. When the fuel is injected into the cylinder, the swirling mixes the air and fuel very well.

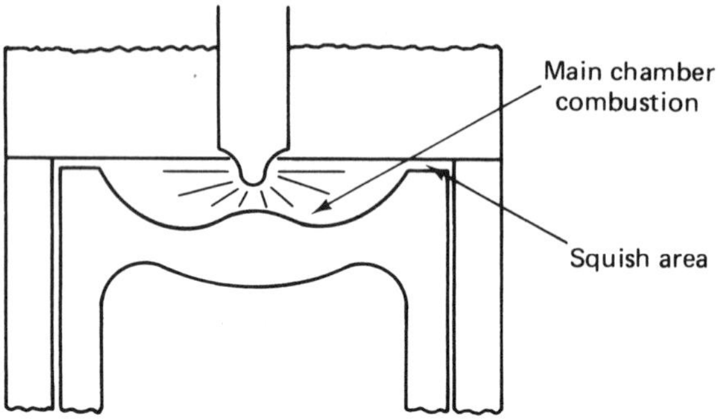

Figure 5-1. The effect of squish area on piston shape.

5.2 Piston and Cylinder Shape

Figure 5-1 shows a piston shape that may be used, in combination with the other factors, to improve combustion. The squish areas are the raised part of the piston crown that are nearest the head. As the piston approaches top dead center, the hot air is forced out of the squish areas back into the main combustion chamber. All two-stroke cycle engines have main chamber combustion, as do four-stroke cycle engines with bores greater than twelve inches.

A turbulence chamber, shown in Figure 5-2, is used on some engines. The turbulence chamber, the circular chamber by the side of the piston, contains 80 to 90% of the piston and clearance volume. Because of the shape of the turbulence chamber, the air is given a swirling motion during the combustion process. After the fuel is injected and ignited the burning fuel and air expand rapidly into the main chamber, where the combustion is completed. Most of the combustion, however, takes place in the turbulence chamber.

5.3 Fuel Injection Nozzles

A different type of fuel injection nozzle, a pintle nozzle (shown in Figure 5-3), is required in a turbulence chamber, rather than the multiorifice type used in main chamber combustion. The multiorifice nozzle (also called a

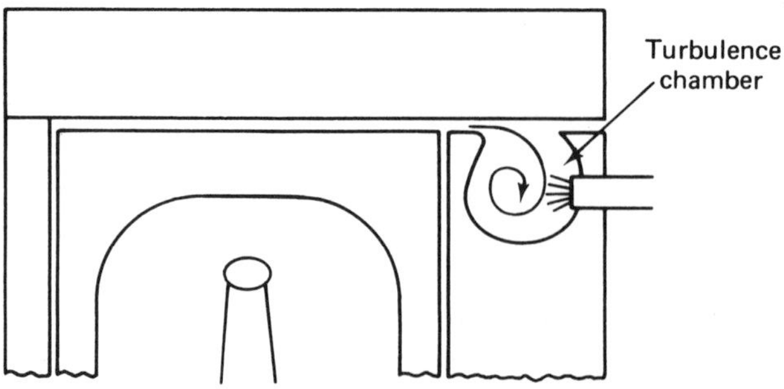

Figure 5-2. A turbulence chamber.

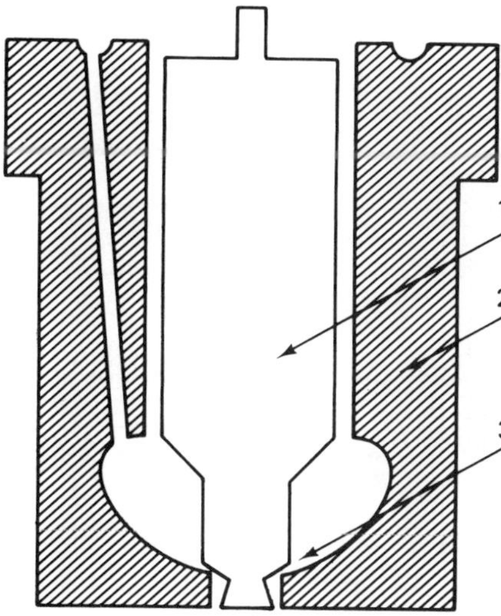

Figure 5-3. Pintle nozzle.

differential hydraulic injector and illustrated in Figure 5-4) requires a fuel oil pressure of 2,500 to 3,500 psi to open, while the pintle nozzle requires a much lower opening pressure. The pintle nozzle opens between 1,200 to 2,000 psi and the fuel sprays in a fine hollow cone pattern rather than a dispersed droplet pattern. There are two requirements for good fuel/air mixing: dispersion and penetration. The multiorifice nozzle does not provide the needed fuel penetration in certain instances. This results in only the air supply near the nozzle being used for combustion, rather than the air supply throughout the combustion chamber. In the turbulence chamber it is desirable that the fuel penetrate across the entire chamber, but it should not impinge the wall of the turbulence chamber.

A modification to the pintle nozzle called a throttling pintle injector is shown in Figure 5-5. The pintle shape is slightly expanded and tapered which causes throttling of the fuel leaving the tip. The fuel spray pattern is conical as in the pintle nozzle and, in addition, a solid stream of atomized fuel is directed on the inside. This causes a penetration across the entire chamber. There are several instances where this type of nozzle is used in the various combustion chambers discussed in this chapter.

5.4 Combustion Chamber Types

The precombustion chamber illustrated in Figure 5-6 (a) uses a pintle nozzle. However, this type of chamber contains only 35 to 45% of the clearance volume. Ignition starts in the precombustion chamber, and the flame and fuel expand rapidly into the main combustion chamber where

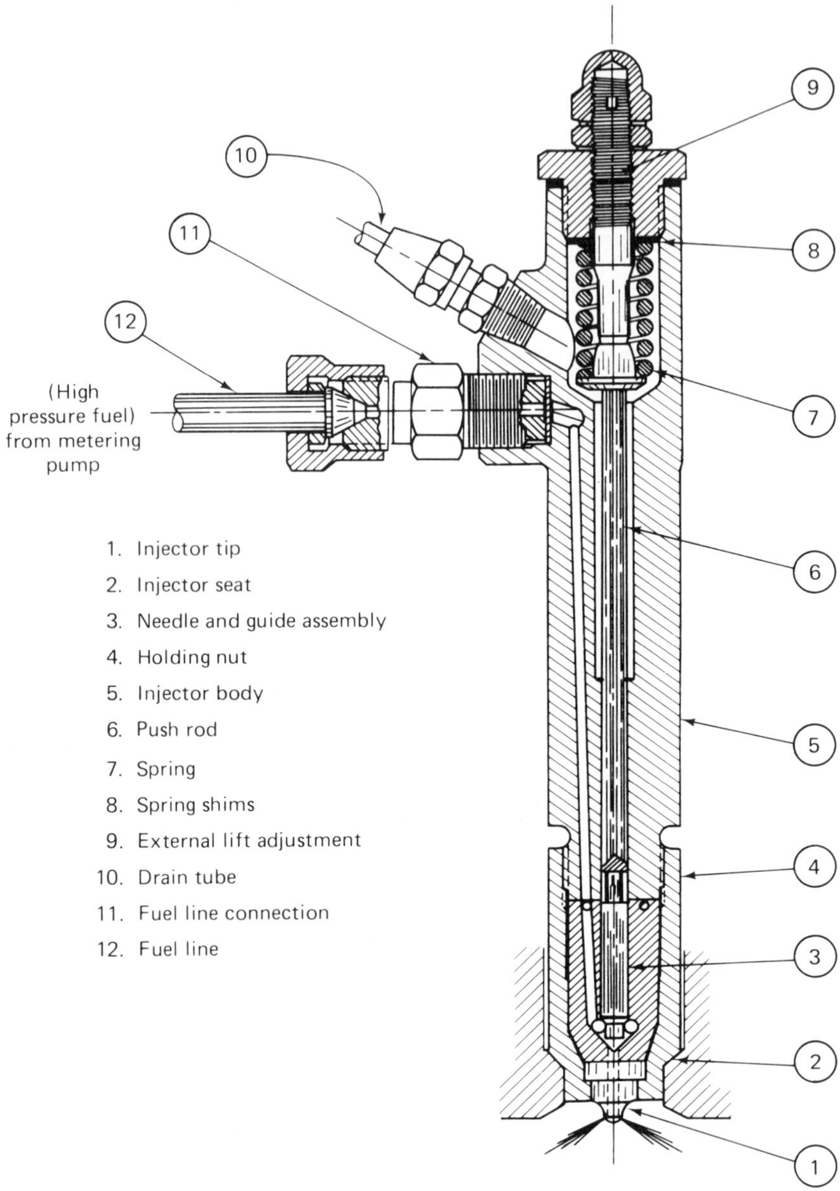

Figure 5-4. Multiorifice nozzle (differential hydraulic injector).

the combustion process continues. An advantage of the precombustion chamber is easier starting, compared with main chamber combustion or turbulence chambers, because the cooling effect of the water jacket on air charge is reduced. The combustion process is also smoother compared to the main chamber combustion or turbulence chamber engines.

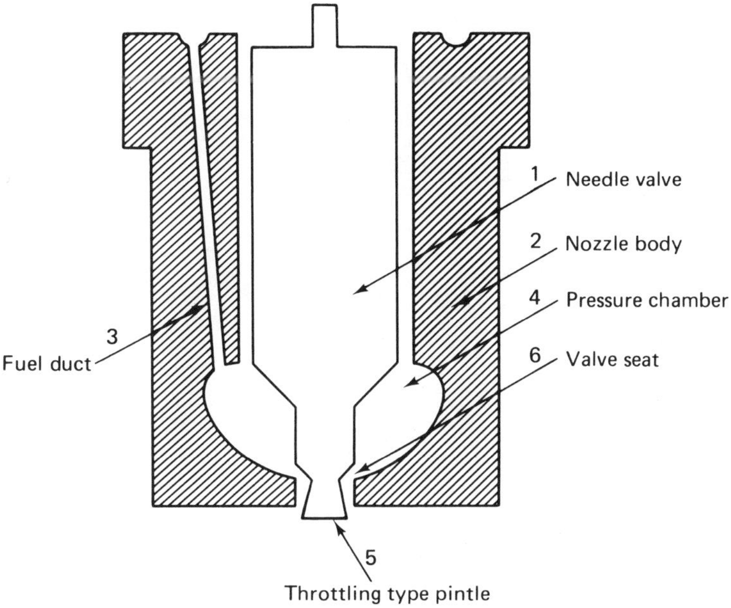

Figure 5-5. Throttling pintle nozzle.

Figure 5-6 (b) illustrates an energy cell. In this case the energy cell only contains about ten percent of the clearance volume. A throttling pintle nozzle is located directly across from the energy cells and a throttling pintle nozzle is used. The fuel is injected across the top of the main chamber where combustion starts, but the solid stream is able to penetrate into the energy cell, and combustion also occurs there. There is a rapid build up of pressure in the energy cell, and the burning gases expand into the main combustion area, creating turbulent mixing of the fuel and air in the combustion chamber. The compression ratio of engines equipped with an energy cell might be as low as 13 to 1. In cold weather such engines would be difficult to start because of low compression pressures and temperature. Located at the entrance of the energy cell is a valve much like a plug cock; the valve should be closed during starting, thus raising the compression ratio from 13 to 1 to about 15 to 1. Once the engine is running and warm, the energy cell is opened and a more efficient combustion process is achieved.

Figure 5-6 (c) illustrates a Lanova energy cell system which is a patented energy cell. The idea is to create as much turbulence as possible in the main combustion chamber. The top view, illustrated in Figure 5-6 (c), shows a cutaway of the main combustion chamber, with the swirl patterns that are typically created by this energy cell.

Finally, there is the air cell, which may be located in a variety of places in the piston or cylinder head. Figure 5-6 (d) illustrates a typical

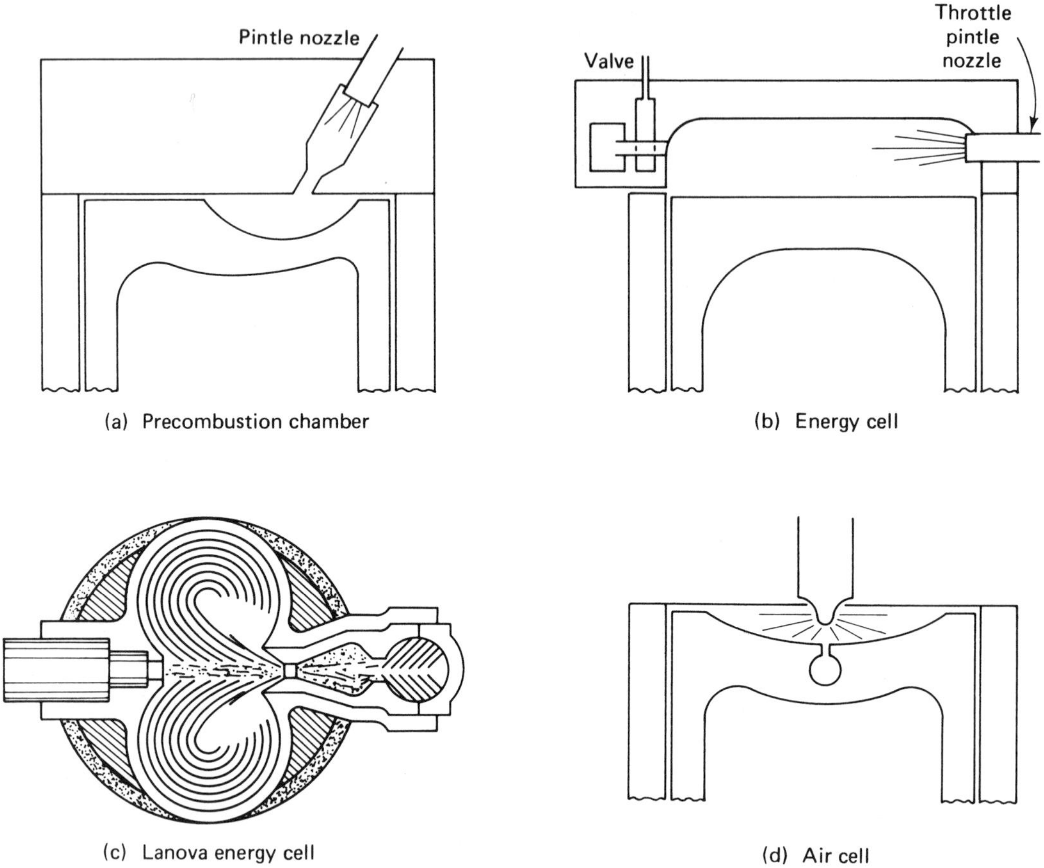

Figure 5-6. Four different types of combustion chambers.

air cell. In the air cell there is just air; no combustion process occurs there. The air in the cell reaches high pressures through compression by peak combustion pressure waves. The downward stroke reduces the cylinder pressure, causing the hot air to leave the air cell and mix with the burning fuel, improving the combustion process and causing still further mixing of the air and fuel.

5.5 Air Intake Systems

The air intake systems on diesel engines have four purposes

1. To supply air for the combustion process.
2. To clean the air before it enters the engine.

3. To silence the air supply so that no whistling or loud noises caused by the air flow through the pipes and manifold are audible.
4. To supply additional air by supercharging the engine.

Paper air filters, similar to those found in an automotive engine, are used on small diesel engines. The cover of the air filter has a felt interliner which reduces noise by damping out vibrations.

Figure 5-7 shows an oil bath type air filter. In filters of this type the air changes direction, but the dirt tends not to change direction and thus remains in the oil. The air also picks up minute amounts of oil which, as the air passes through the wire mesh strainer, clings to the wire, further removing any dirt particles that may have been carried along. The oil picked up by the mesh drips back down into the sump of the filter. There is a danger in overfilling this type of oil filter in that the oil may carry over into the intake system. In this case the engine will be running on lube oil rather than diesel oil and it will run without governor control. The governor can control only the fuel oil; it cannot control the lube oil. If there was a serious overflow, there could be a runaway engine.

Air silencers are illustrated in Figures 5-8 and 5-9. In Figure 5-8 a slotted pipe is used, the silencer is designed to break up the frequency of the air, allowing no harmonic vibrations, which cause loud noises. In this case the minimum area of the slots has to be $1\frac{1}{2}$ times the pipe inlet area.

Figure 5-9 illustrates an acoustic silencer. The operational principle is the same; the air frequency is disturbed, preventing harmonics. These units use a variety of holes, rather than slots, to break up the air harmonics. Acoustic silencers are sometimes used in combination with regular air silencers. In that case, they mount under the regular unit.

Another type of air cleaner is the direct impingement unit. These

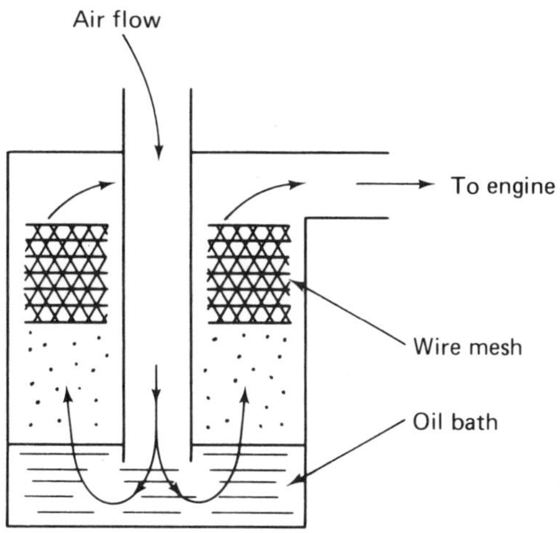

Figure 5-7. Oil bath type air filter.

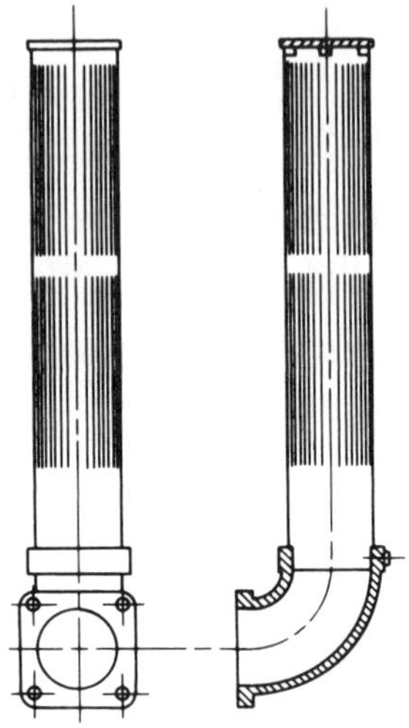

Figure 5-8. Air silencer.

typically consist of a cubical shape whose four sides are framed and enclosed with 1/4 to 1/2 in. thick steel wool or wire mesh. The bottom and top are solid. The wire mesh is dipped in lube oil and the air flow moves through the wire mesh, causing the dirt to cling to the lube oil. Periodically, these filters must be taken out and washed. They are washed in fuel oil and blown dry so that there will be no fuel oil in the air as it goes to the intake manifold. Before being reinstalled, the filters are dipped in lubricating oil, allowed to drain so the excess oil is removed, and then replaced in the filter frame.

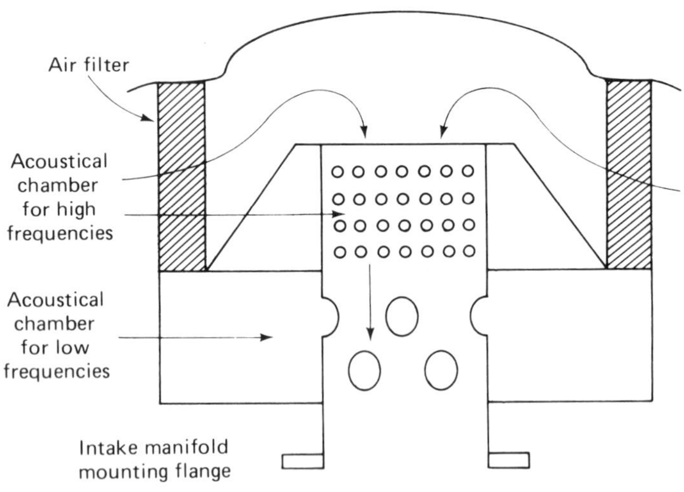

Figure 5-9. Acoustic air silencer.

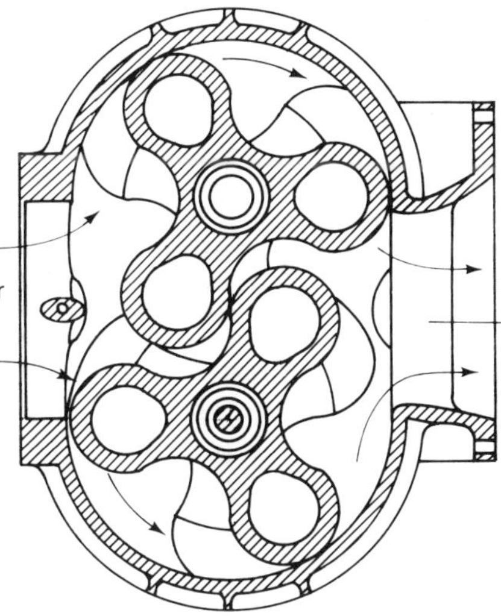

Figure 5-10. Roots-type blower.

5.6 Air Intake Systems for Two-Stroke Cycle Engines

All two-stroke cycle engines require some means for scavenging air. The air must be at greater than atmospheric pressure to force the exhaust gases out of the cylinders. On smaller engines a Roots-type blower might be used as shown in Figure 5-10. There are driving gears on one end of the blower and either a lobe-type compressor or a positive displacement pump on the other end. Lobes force the air through the pump; it is forced out on the discharge side at a higher pressure. The gears maintain the space between the lobes so there is no metal-to-metal contact. On some engines, double acting piston pumps are used.

Figure 5-11 shows another practice, called underpiston scavenging, that has been used. This technique may be used only on crosshead en-

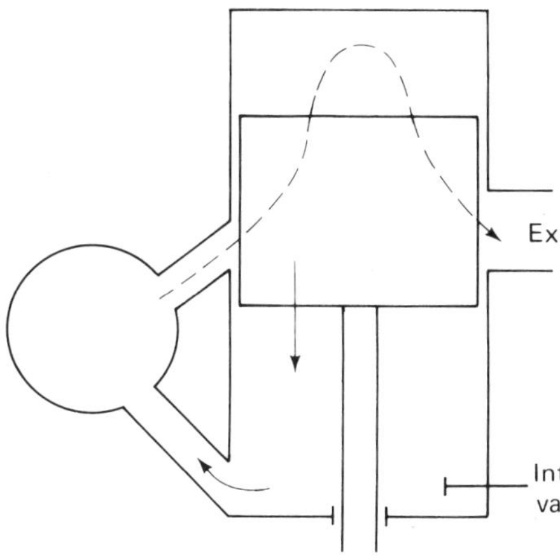

Figure 5-11. Underpiston scavenging.

gines because the piston rod is connected to the crosshead and thus the piston and the piston rod move in the same vertical plane. The inlet valves are in the lower part of the cylinder and the downward motion of the piston compresses the air, which enters an intake receiver. When the intake port is uncovered by the crown of the piston, the air in the receiver enters the main cylinder and forces out the exhaust gases. On the upward stroke the inlet valves in the scavenging section open and new atmospheric air enters. The rotary valve blocks the exhaust port and prevents air loss. Underpiston scavenging is currently used on many large diesel engines. Rotary exhaust valves, however, have been eliminated from current engine design because high temperatures and carbon build up erode the the exhaust valve seals rapidly. Other rotary valve problems include the difficulty in timing the valves and replacing worn parts.

Figure 5-12 illustrates a problem common to ported two-stroke cycle engines. The exhaust port must open first so the hot exhaust gases can escape, reducing the pressure in the cylinder before the intake port opens. This means that on the upward stroke the intake port is going to close before the exhaust port. This allows a reduction in air charge in the cylinder as some of the new charge exits through the exhaust port. If at all possible, the exhaust and intake ports should close simultaneously.

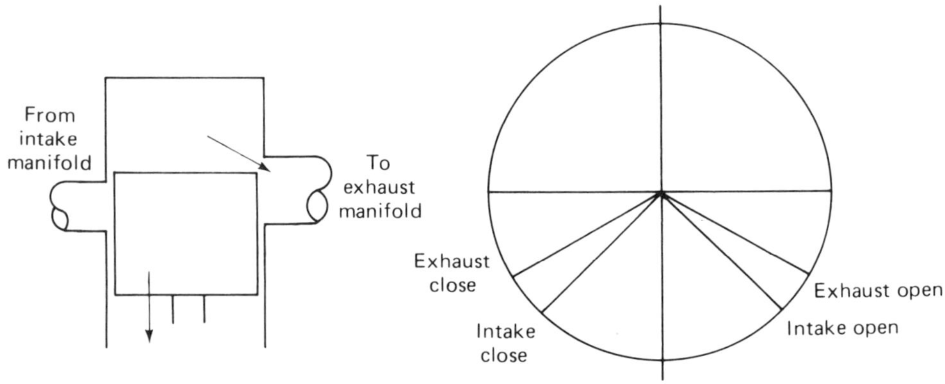

Figure 5-12. Ported two-stroke cycle engine.

Figure 5-13 shows two methods for achieving simultaneous closure. In Figure 5-13(a), nonreturn flow valves are located on the intake manifold. Sometimes these valves are called supercharging valves. These valves allow the intake port to open first and close later, but will not allow exhaust gases into the intake manifold. Gases can only flow from the intake manifold into the cylinder. Hence, the air charge in the cylinder may be increased. In Figure 5-13(b), a rotary exhaust valve closes the exhaust port on the upward motion of the cylinder at the same time the intake port is closed by the piston.

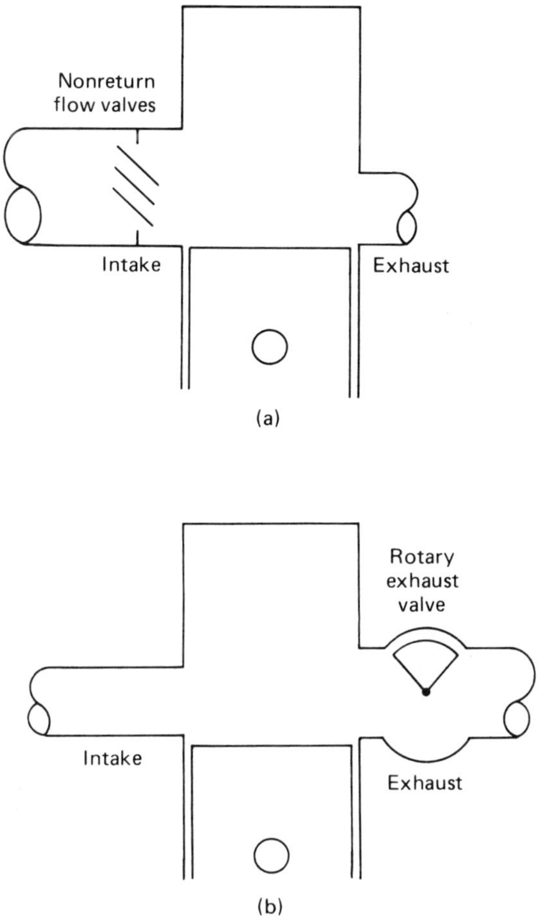

Figure 5-13. Methods for achieving simultaneous exhaust and intake port closure. (a) Nonreturn flow valves; and (b) Rotary exhaust valves.

5.7 Turbocharging

It is possible to use the energy of the exhaust gases to drive a turbine and then have this turbine drive a compressor or, as it is often called, a blower. The combination is called a turbocharger, and it operates in conjunction with positive displacement methods of scavenging air on an engine.

Figure 5-14 shows an example of the parallel operation of a piston scavenging pump on a crosshead engine and a turbocharger. In slow speed operation, until the engine reaches about 50% of its rated load, positive displacement methods of supplying supercharging or scavenging air must be provided as the turbocharger will not provide sufficient amounts of air. An underside scavenging air pump is used as shown in this figure. There is a check valve in the manifold that runs from the blower to the engine to prevent air from going to the blowers from the air pump at

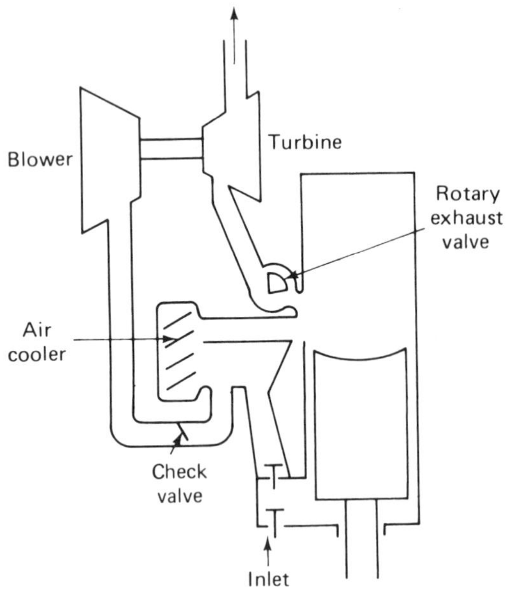

Figure 5-14. Parallel operation of turbocharger and piston scavenging pump.

slow speed operation. Notice that the air is cooled before it enters the engine. The cool air has a greater density, and hence greater mass entering the engine. The heat exchanger is typically salt water cooled.

Figure 5-15 illustrates a series-type scavenging system where the suction is through the turbocharger-driven blower at all times. There are nonreturn flow valves in this type of engine and the air flow is from the turbocharger, through the reciprocating scavenging pump, into the header. The nonreturn flow valves in the header prevent the backflow of combustion gas into the intake manifold.

In Figure 5-16, a means of turbocharging at high loads with parallel operation at low loads (speeds) is shown. When the engine reaches about 40 percent load (as measured by governor setting), the clutch disengages the Roots-type blower and all air is supplied by the turbocharger. As the engine slows down (for instance, during maneuvering, to 30 percent load) the clutch reengages the Roots-type blower and parallel operation occurs. On the newer General Motors (EMD) engines, the turbocharger is driven by the engine gear train during starting. The gear ratio is stepped up so the turbocharger rotates at high speeds and is able to supply scavenging air during the start period. An overrunning clutch allows the turbocharger to run faster than the gearing allows once the engine reaches a load (speed) where the thermal energy of the exhaust is great enough to cause it do so. Should the turbine slow down, which means the exhaust gas temperature and volume flow rate are not great enough to provide the required air, the clutch reengages and the turbine is again gear driven.

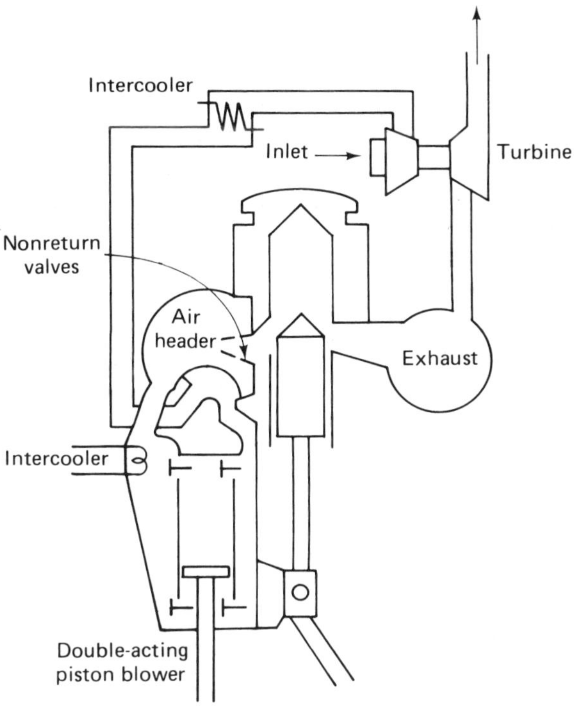

Figure 5-15. Series operation of a turbocharger and pump.

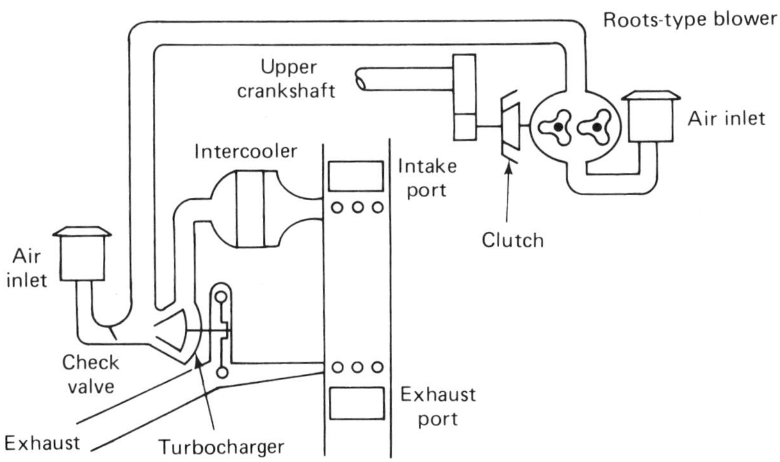

Figure 5-16. Overrunning clutch operation.

5.8 Four-Stroke Cycle Air Intake Systems

Four-stroke cycle diesel engines are normally self aspirated, which means that they do not require a mechanical air supply system. The piston will provide the necessary air supply on its intake stroke. What is needed is an air cleaner, an intake manifold, and the inlet valves in the head. However, it is often desirable to get more horsepower by supercharging the engine. Supercharging allows a greater mass of air; thus, a greater mass of fuel can be burned in the cylinder, providing a greater release, increasing the power potential.

There are four advantages to turbocharging. The first is to increase the mass of air in the cylinder; the second is to increase the brake mean effective pressure; the third is to increase the brake horsepower and the fourth is to decrease brake specific fuel consumption. Increasing the brake mean effective pressure takes into account the design efficiency of the engine. The brake mean effective pressure is a criterion that is used to compare different engines at different operating conditions. The indicated mean effective pressure may not necessarily be higher between different engines even though bmep's are different. The indicated mean effective pressure may actually be higher on an engine that has a lower brake mean effective pressure. The brake mean effective pressure takes into account the mechanical design, the efficiency with which the power is transmitted to the shaft. Turbocharging may increase the brake horsepower 50 to 75% over the same nonturbocharged engine. A 1,000 horsepower self aspirated engine might produce 1,500 to 1,700 horsepower when supercharged.

Two results of supercharging are that the exhaust temperature is not significantly raised and that the brake specific fuel consumption drops as the fuel required for each horsepower produced is less. Nevertheless, since the horsepower output is increasing, the total fuel required increases but the brake specific fuel requirement decreases.

The cooling system on a self aspirated engine may or may not have to be changed to accommodate turbocharging. The total peak pressure of the engine is not significantly higher in the turbocharged engine.

Figure 5-17 illustrates the *p-V* diagram for a properly supercharged engine. The effects of higher peak pressures, if precautions are not taken, are superimposed on a *p-V*diagram for a properly supercharged engine. If precautions are not taken when a self aspirated engine is supercharged, it is possible to overload the engine. The peak pressure and temperature may be very high, causing bearing and piston failure. Normally the compression ratio is lowered as a precaution; if the compression ratio were 15 to 1 for a self aspirated engine, it would be lowered to 13 to 1 for a supercharged engine.

The cooling system has to remove about 20 percent of the heat produced by the engine. If the amount of fuel burned is increased, then the

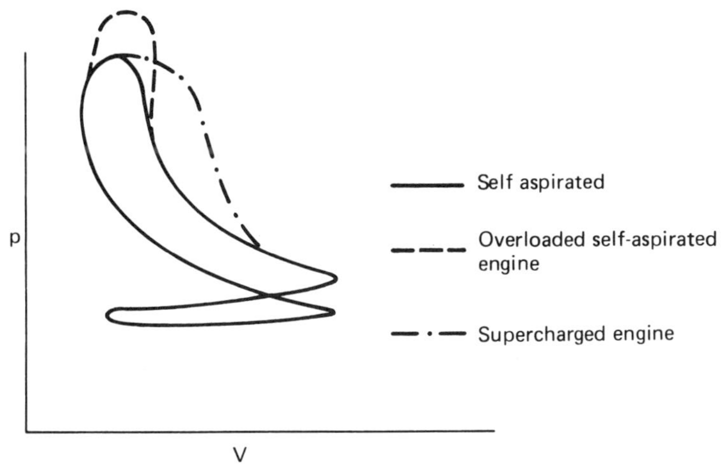

Figure 5-17. Effect of supercharging on four-stroke cycle *p-V* diagram.

total heat supplied to the engine is increased, and the total heat removal responsibility of the cooling system is also increased. For example, if the power is increased 30%, 30% more cooling would be required. There is a way that the supercharged engine is able to bypass some of this cooling responsibility. Figure 5-18 illustrates the polar timing diagram for self aspirated engines with the dotted lines superimposing the location of the intake valve openings and the exhaust valve closings for a turbocharged engine. Notice that there is an overlap now of almost 120° rather than the 10–40° previously shown. This extended overlap period allows cooling of the exhaust valves, piston head, and cylinder walls by the air system; hence, some of the cooling responsibility that might fall on the cooling system is taken care of by the cooling effect of the air. It may be necessary to provide the engine with new circulating water pumps and/or water jacket volume. Had a self aspirated engine been converted, a new engine camshaft would be required so that the valves would open at the proper time.

As noted earlier in the discussion of the parallel operation of turbochargers, intercoolers are used in conjunction with turbochargers. This is because the compression process increases the temperature and volume flow rate of the air. The volume flow rate is inversely affected by the density, and it is desirable to have the air temperature entering the cylinder as low as possible, allowing a greater density, and thus a greater air mass, in the cylinder. The intercoolers are salt water cooled because the fresh water available for cooling is at too high a temperature. It is desirable to cool air to 100°F (38°C) if possible, and the fresh water cooling system is operating at 160–180°F (70–80°C).

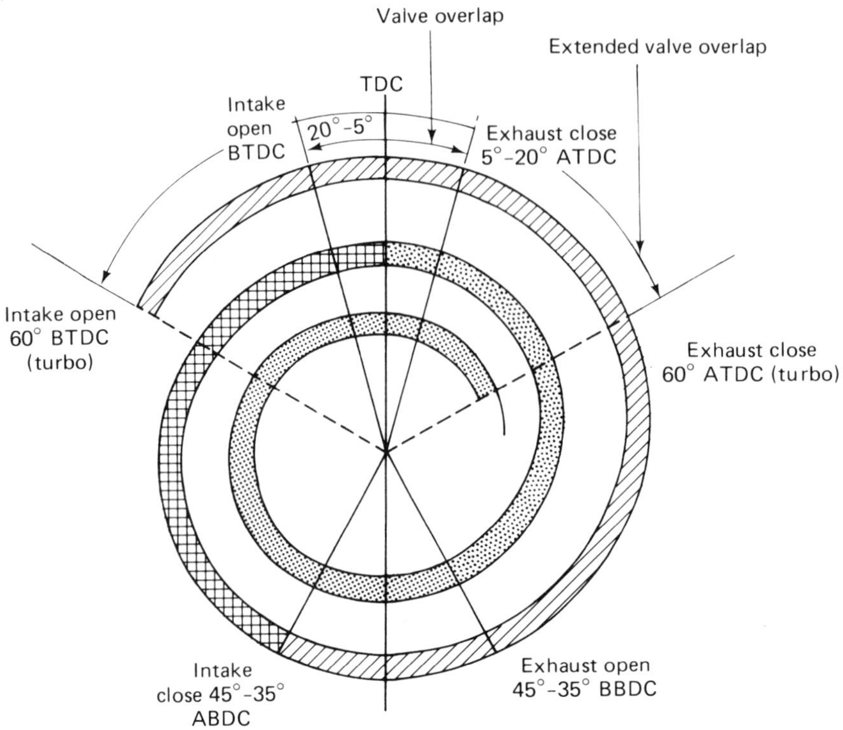

Figure 5-18. Differences between self aspirated and supercharged engines.

Supercharging causes higher average pressures to act on the bearings. These are not necessarily higher peak pressures but higher sustained bearing pressures, which are illustrated by the increased work (the increased area) in the *p-V* diagram in Figure 5-17. This increase in pressures requires that there be an increase in oil flow to maintain the proper bearing oil temperature. If the same oil flow were maintained, the greater force on the bearing would cause the oil temperature to rise, increasing the probability of bearing failure. Thus, new lube oil pumps are required. Also, new fuel injection pumps would be required, having the capacity to provide the total amount of fuel necessary for the turbocharged engine.

Should a decision be made to use a Roots-type blower, keep in mind that the Roots-type blower speed is directly proportional to the engine speed; it is a positive displacement compressor. Hence, if the engine overloads by burning more fuel, there will not be more air. The engine speed will not increase with load; it may decrease a little. This means that the exhaust temperature will rise, increasing the potential for burned exhaust valves and serious damage to the piston and cylinder due to thermal stresses. Turbocharging, on the other hand, increases the air supply with load. This is because the exhaust gases directly provide energy for the compression process. As the engine is loaded, the exhaust temperature increases, providing more energy to be extracted in the turbine and hence, more air which may be compressed.

5.9 Exhaust Systems

The primary purpose of the exhaust system is to remove the exhaust gases of the engine. In addition, the exhaust system silences the noise of the exhaust gases, quenches sparks, and furnishes energy for turbochargers and/or waste heat boilers. On the four-stroke cycle engine, the exhaust system begins with the exhaust valve, continues with the exhaust manifold, and at the end of the engine is attached via a flexible manifold to the uptake. At the top of the uptake, a spark arrester and a silencer are attached, eliminating sparks and reducing the noise of the exhaust gas. The gases then go into the tailpipe, discharging to the atmosphere. The mufflers, i.e., silencers, which are most typically used are the frequency break up type, similar to those on the intake air system. As before, the frequency is broken up to prevent the harmonics that create noise. Waste heat boilers, if used, may be of the fire tube or water tube type and the fittings on the boilers are similar to those on a standard marine boiler.

6

Fuel and Fuel Injection Systems

6.1 Fuel Classification

Diesel fuels can be classified by their distillate content and their cetane rating. There are five categories of diesel fuel, numbered from one to six, but skipping number three. Number one diesel fuel is a high grade distillate whose primary use is in high speed diesel engines; the fuel has a cetane rating between 40 and 65. It is very clean and clear. Number two diesel fuel is also a distillate fuel used in high and medium speed engines. Its cetane rating is also between 40 and 65. Cetane rating is related to a fuel's ignitability. High speed engines, in particular, require that the fuel ignite quickly as there is a short time for the combustion process. There is no number three diesel fuel. Number four diesel fuel is a blend of distillate and residual fuels. This fuel needs no heating and is used in medium and slow speed diesel engines. Its cetane rating normally ranges between 30 and 45. Number five diesel fuel, also a blend of distillate and residual fuels, is used for slow speed diesel engines, and normally requires heating to meet the viscosity requirements of the fuel injector. Number six diesel fuel is typically a Bunker oil with no cetane rating. This fuel invariably requires heating before it may be used in the injection system.

Aboard large ships there may be two fuel systems, one for light and the other for heavy fuel. The light fuel oil system is common to all ships.

6.2 Fuel Oil Systems

Figure 6-1 illustrates a light fuel system commonly found aboard diesel-powered vessels. The various components in the system are depicted, and the flow through the system is shown. Heavy fuel systems (not shown) are very similar, except that they require a device for heating the oil. Heating coils are located in the tanks and oil heaters follow the fuel booster pumps. Injectors call for a minimum specification of between 75 and 125 SSU, the exact specification depending on the manufacturer. When a heavy fuel system is used, as on most large and medium speed diesel engines, a light fuel system is also necessary. This is called a two fuel system, and it is important when maneuvering and coming into port that the ship is switched over to the light fuel system before arrival. This prevents the build up of heavy fuel residues in the injectors pumps, and piping, avoiding the possibility of heavy fuel oil congealing in the high pressure fuel lines between the fuel oil metering pump and the injector, and in the injector itself. No fuel heating is necessary when using the light fuel system. This allows easy cold starts, impossible with most heavy fuel systems. Furthermore, maneuvering often requires that the engine be stopped and restarted quickly and cleanly, and this calls for fuel with a higher cetane rating than steady running demands.

6.3 Injection System Requirements

All injection systems must meet five basic requirements. First, they must provide accurate metering. Each cylinder's metering fuel pump must provide the exact amount of fuel delivered to its cylinder, as the next fuel pump does to its cylinder. Otherwise, a cylinder could be overloaded, resulting in high peak pressures and temperatures in that cylinder. In addition, uneven fuel distribution will cause the engine to run rough and hunt (hunting is a rhythmic variation in speed). Fuel imbalance to the cylinders may be diagnosed by reading the cylinder exhaust temperatures, which should be within 10°F or 5°C of each other. If they are not, ad-

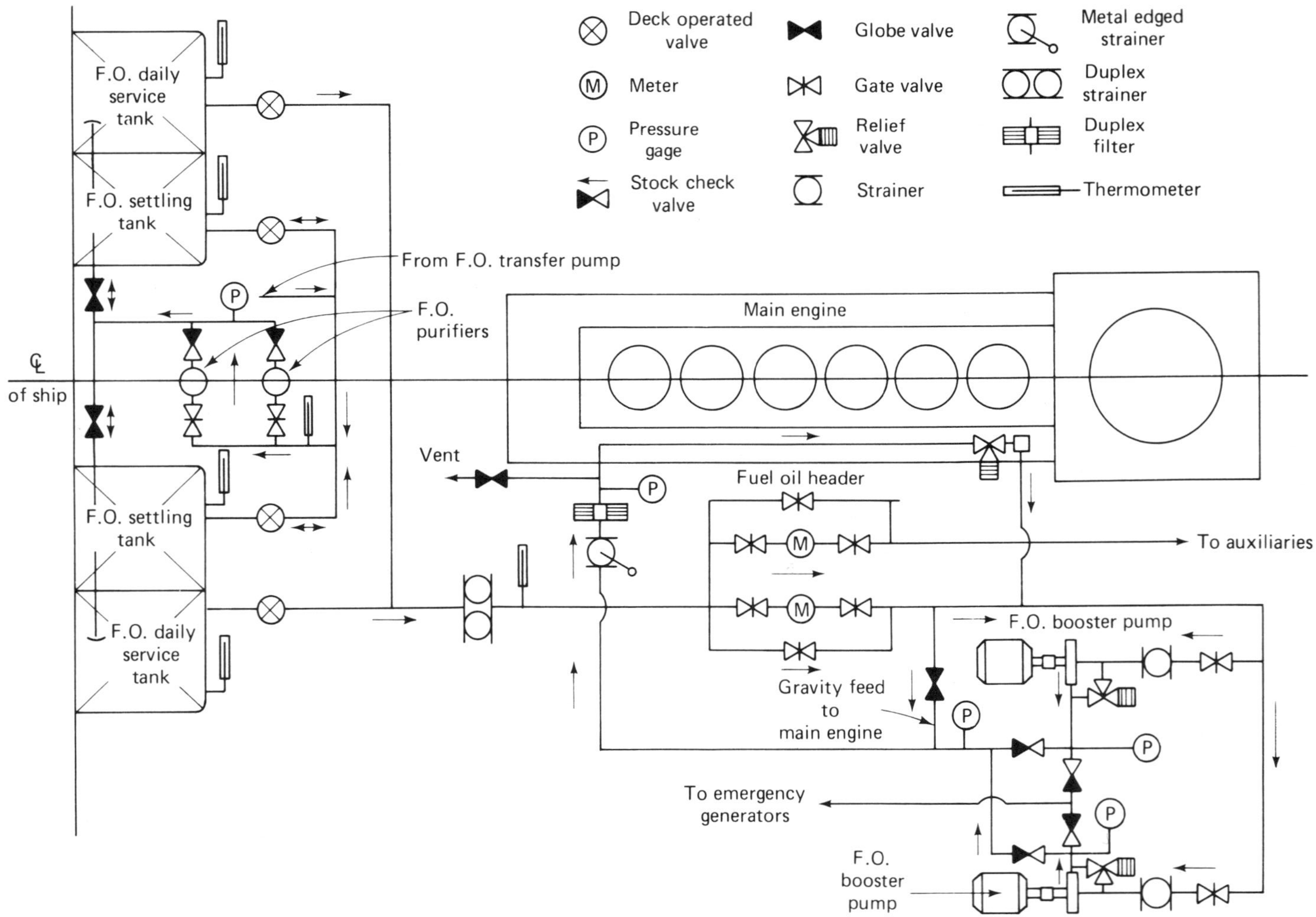

Figure 6-1. Light fuel oil systems.

justments should be made to each fuel pump until the proper balance is attained. Second, there must be proper timing. The time of the beginning of injection in degrees before top dead center must be the same for all cylinders. The time is specified by the engine manufacturer. Third, there must be a suitable rate of injection. The fuel mass per degree of crank travel must be the same for all the fuel pumps. Fourth, the fuel must be properly atomized; that is, broken mechanically into droplets as fine as possible. Fifth, there must be proper penetration of the fuel into the cylinder. This allows combustion to occur over a greater area of the cylinder. A balance must be struck between requirements four and five. If there is too much atomization of the fuel, then the penetration into the cylinder will not be sufficient and combustion will occur in a very small space within the cylinder. This means that the fuel will not mix sufficiently with the oxygen and combustion will be incomplete. If the penetration is too great, then there is not enough atomization. At that other extreme, the fuel may penetrate as a solid stream across the cylinder. Combustion would not be complete, as the unatomized fuel would not mix properly with the oxygen. Furthermore, penetration in a solid stream could result in raw fuel striking the piston crown, which causes erosion and cracking of the piston.

6.4 Injection System Classification

One of the earliest types of fuel injection system was the common rail type, a mechanical system. The inertia of this mechanical system precluded its use on high speed diesel engines, so new and better fuel injection systems were developed. A fuel injection system may be classified according to how it begins and ends injection. There are four combinations. The first is a variable beginning with a constant ending; second, a constant beginning with a variable ending; third, a variable beginning with a variable ending; and, fourth, a constant beginning with a constant ending. These four types allow for variations in engine loading and timing. For instance, constant speed machines, such as diesel-driven generators, may require that the starting of injection be the same but that the ending be variable to account for load changes.

6.5 Metering Pumps

Fuel injection pumps are classified by the mechanism they use for metering. They fall into two categories: the first is the variable stroke metering pump, where the actual stroke of the metering pump varies in length of travel, resulting in more fuel delivered at full load (longer stroke) than at low loads (shorter stroke); the second is the variable effective stroke pump, where the actual pump stroke is constant, but the fuel delivered by the pump may be varied.

A wide variety of different designs of metering pumps have been used on diesel engines; some of those encountered are illustrated in Figure 6-2. These pumps use different metering principles.

The swinging arc pump, shown in Figure 6-2(a), is classified as a variable stroke pump with constant beginning and constant ending of injection. When the roller, which is attached to the pump plunger, is in position 1, the plunger displacement will be at its maximum. In position 2 the plunger will have no vertical movement at all, resulting in zero fuel

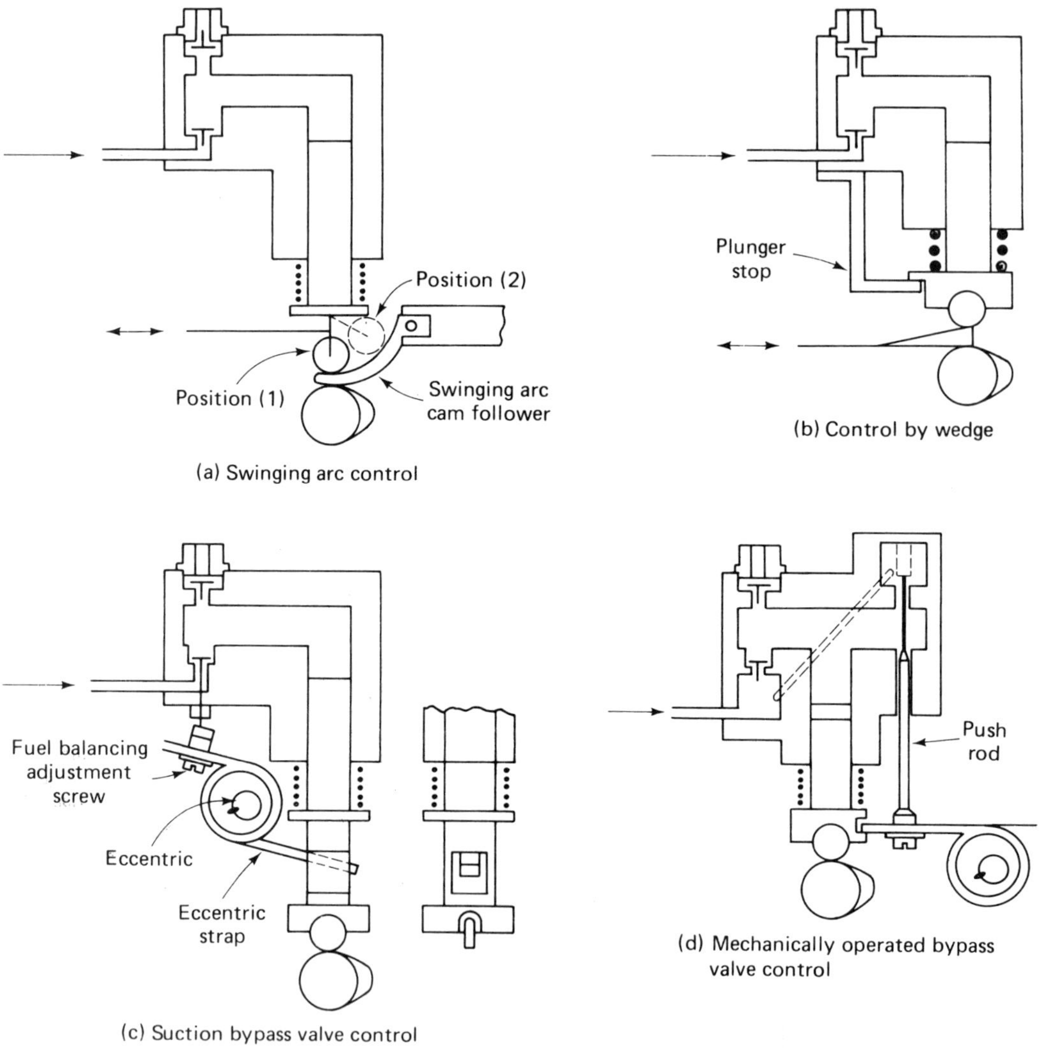

Figure 6-2. Types of fuel metering pumps.

delivery to the engine. Thus, the position of the swinging arc cam follower determines the pump stroke length, and variation in fuel flow is achieved. Note that a requirement of all diesel engine fuel pumps is that they have a position of zero fuel output. This is to allow the operator to stop the engine.

Figure 6-2 (b) illustrates another type of variable stroke pump, in which a wedge is used as a cam follower. Again, the stroke may be varied from maximum to zero by the wedge position. A plunger stop holds the plunger in the barrel at the zero load setting. The wedge is positioned by the engine governor. On this system the ending of injection is constant, but the beginning is variable, depending on the clearance between the plunger and the wedge.

Figure 6-2 (c) illustrates a suction bypass pump, where the stroke is constant, but the fuel displaced is varied. This is accomplished as follows: an eccentric is attached by a strap to the pump plunger at one end and contacts the pump suction valves at the other end. The eccentric holds the suction valve off its seat, resulting in no pressure build up, and hence, no injection. The plunger always moves the same distance, but the opening of the suction valves determines the amount of fuel delivered. The governor controls the eccentric position. This pump has a variable beginning and a variable ending.

Figure 6-2 (d) illustrates a mechanically operated bypass valve pump, which is similar in operation to the suction bypass pump. There are three valves in this pump: suction, discharge, and bypass. The bypass valve allows a flow of high pressure fuel to the suction side of the pump. The pushrod will contact the bypass valve on its vertical rise, causing it to open and stopping fuel discharge to the injector. The eccentric position, determined by the governor, allows the ending of injection to vary. The beginning is constant.

Figure 6-3 illustrates a combination of a suction bypass and a spill valve pump. There is a suction valve operated by an eccentric, which in turn is controlled by the governor. The effective stroke of the pump is controlled by the setting of the suction valve closing. The spill valve controls the ending of the stroke, which is constant; it is actuated by another eccentric. In this case the fulcrum is fixed, so the ending does not vary. This type of pump is found on two-stroke cycle engines.

The rotary plunger type fuel injection pump is the most common, probably accounting for 90% of the market. Figure 6-4 illustrates the plunger of a rotary plunger pump. The plunger is the key to the pump operation. Notice that a helix is formed on the pump plunger and that the left and right hand sides have ports which are opened to fuel supply coming from the booster pump. When the plunger is below the ports, fuel from the booster pumps (at comparatively low pressure) is everywhere throughout the system. It is below and above the helix area. A

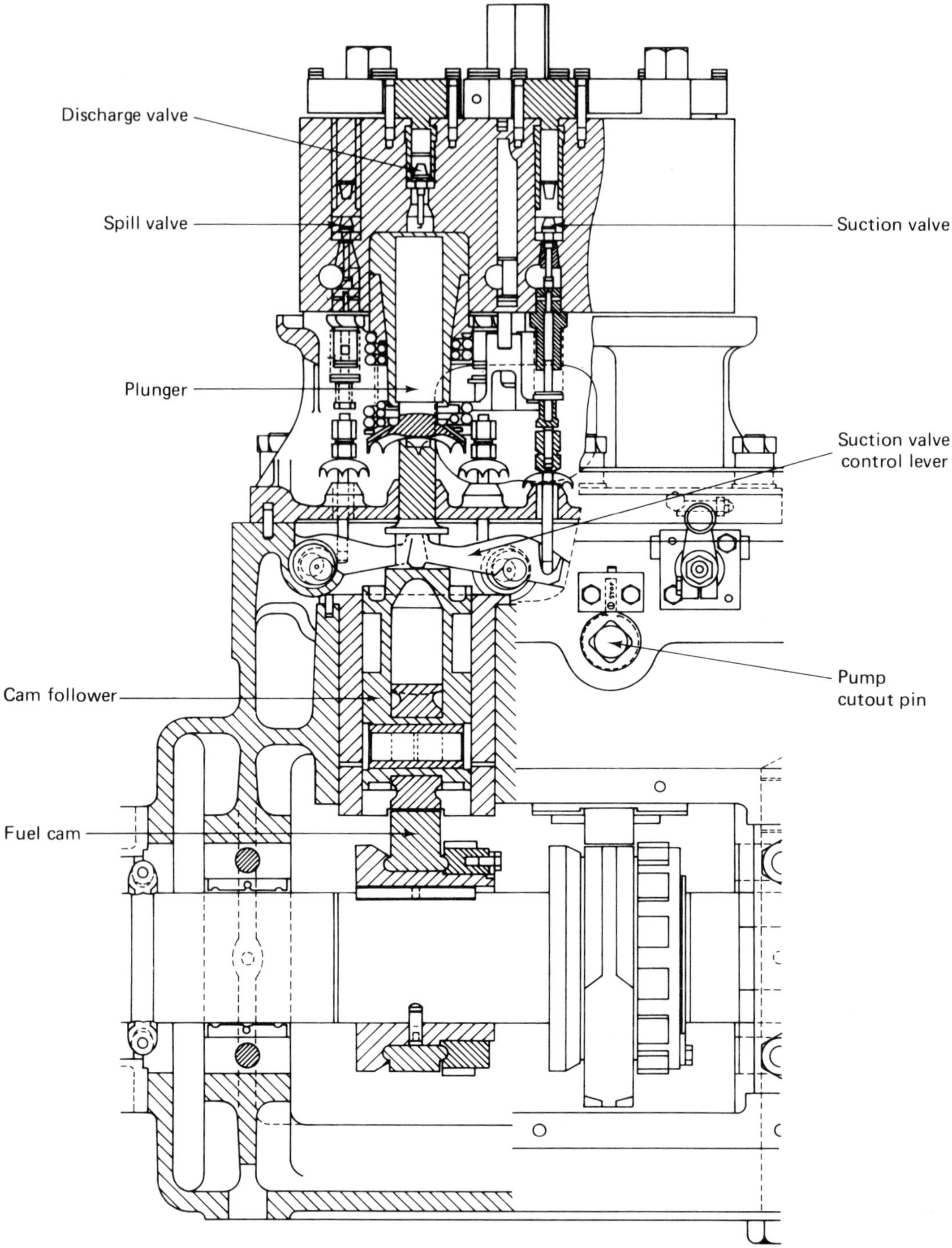

Figure 6-3. Combination suction bypass and spill valve pump. (Reproduced with permission from Sulzer Bros., Inc.)

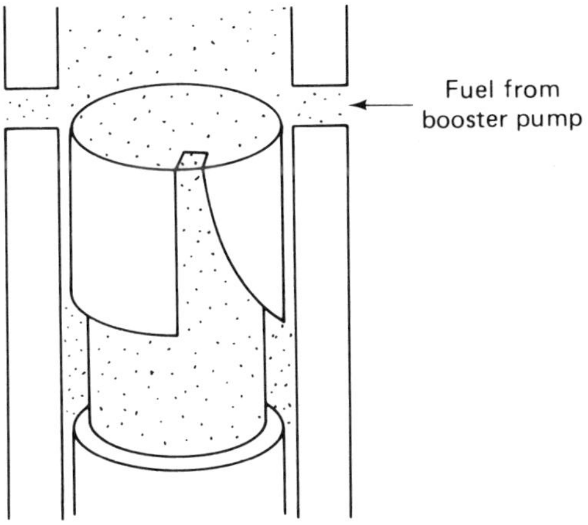

Figure 6-4. Rotary plunger pump.

very tight fit between the plunger and the barrel is required, as the fuel oil pressure may reach 3500 psi during the injection process and there can be no leakage. As the plunger moves vertically upwards, the ports are closed and injection begins. This is illustrated in Figure 6-5. Referring to Figure 6-5, as the ports close, the beginning of injection starts. No more fuel is allowed in. The plunger proceeds upward, increasing the pressure, and the fuel is forced through the injector and into the engine itself. As the helix moves up and uncovers the port, the fuel flows from a high pressure region to the low pressure supply. This is called the end of injection. Thus, the pump has a constant beginning and a variable ending, depending on where the helix uncovers the port.

Fast plunger acceleration during the beginning of injection is desirable so that the pressure builds up quickly, allowing the needle valve in

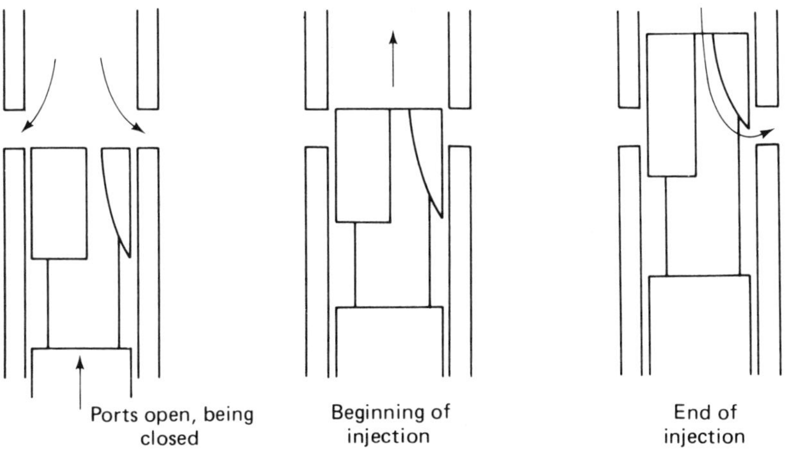

Figure 6-5. Injection sequence for a rotary plunger pump.

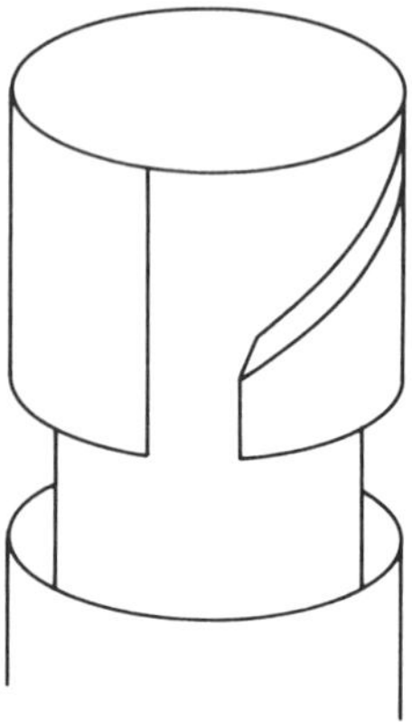

Figure 6-6. Variable beginning, constant ending plunger.

the injector to snap from its seat, thus preventing the dribbling of fuel from the injector tip. If the fuel pressure was allowed to increase slowly, injection would be delayed and would occur at too low a pressure. It is also desirable that the pressure decrease quickly at the end of injection. This prevents the fuel from tapering off and dribbling at the nozzle, which could result in incomplete combustion as well as coking, the formation of carbon deposits on the nozzles. When coking occurs the fuel does not burn completely, but solidifies in the small passages, and forms carbon cones on the outer surface of the tip. This carbon formation distorts the spray pattern of the nozzle, causing improper combustion, possibly black smoke, and possible fuel impingement on the piston crown.

The fuel cam is shaped to control the plunger accelerations and decelerations. The plunger in this type of rotary plunger pump is governor controlled. The barrel is fixed, but the governor rotates the plunger, regulating the cutoff of the fuel.

Figure 6-6 illustrates a variable beginning, constant ending plunger. All plunger pumps are timed to some constant value. In the previous case (Figure 6-5), the constant value was the port closing; the port always closed at the same time, regardless of how the plunger was rotated. This time (port closing) is the beginning of injection. In Figure 6-6, port opening is constant for all rotations of the plunger, so the timing of this type of fuel injection pump is for port opening or the end of injection. About 90% of the rotary plunger pumps have a constant beginning of injection.

6.6 Effect of Pressure Waves

Figure 6-7 shows a simplified metering pump/fuel injector setup on a diesel engine. Assume that the pressure required to open the injector is 3500 psi, so the metering pump will compress the fuel until the pressure

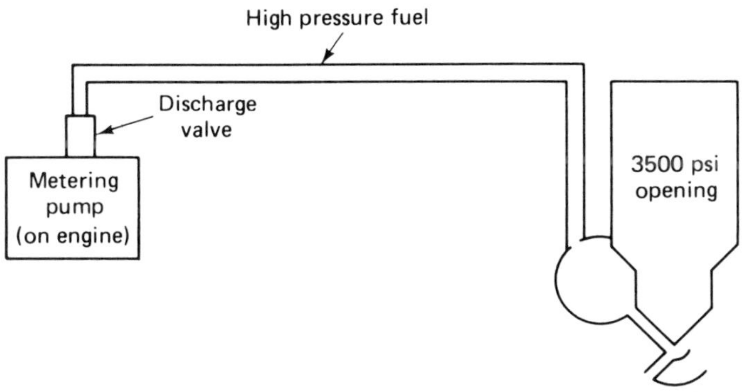

Figure 6-7. Simplification of the high pressure side of fuel system.

reaches 3500 psi. Once the injector opens, there is a greater surface area and, perhaps, only 3200 psi is necessary to keep the injector open. Because of the pulsating motion, a 400 to 500 psi shock wave moves back and forth in the fuel line connecting the meter pump to the injector. Assume that injection ends, the valve seats, and the pressure acting on the injector is 3200 psi. When the reflected shock wave comes back to the injector, another 400 to 500 psi is added to the fuel pressure, so now there is between 3600 and 3700 psi at the injector. It opens again, allowing fuel to enter the combustion chamber, and combustion continues. This is called after-injection and must not be allowed. To prevent this, a delivery, or discharge, valve is installed in the metering pump. The delivery valve has a retraction piston that allows expansion of the fuel oil by occupying a certain volume in the high pressure line connecting the metering pump to the injector. When the metering pump port opening occurs, allowing the fuel pressure to drop, the retraction piston volume is available for the fuel, causing a rapid decline in the pressure in the fuel line between the injector and the pumps. Hence, even though the shock wave has been moving in the fuel line, the pressure drops significantly below the 3200 psi threshold. The size of the retraction piston must be determined for each system as it depends on the volume of fuel contained in the line going from the metering pump to the injector. Fuel has about a one percent compressibility by volume at a pressure of 3000 psi. The retraction piston must also be sized to account for this compressibility. A problem may occur aboard ship if discharge lines rupture and are replaced with lines of a different length, particularly if they are longer, because then the volume that the delivery valve occupies may not be sufficient to prevent after-injection. Care must be taken to insure that replacement fuel lines are the same length as the original units.

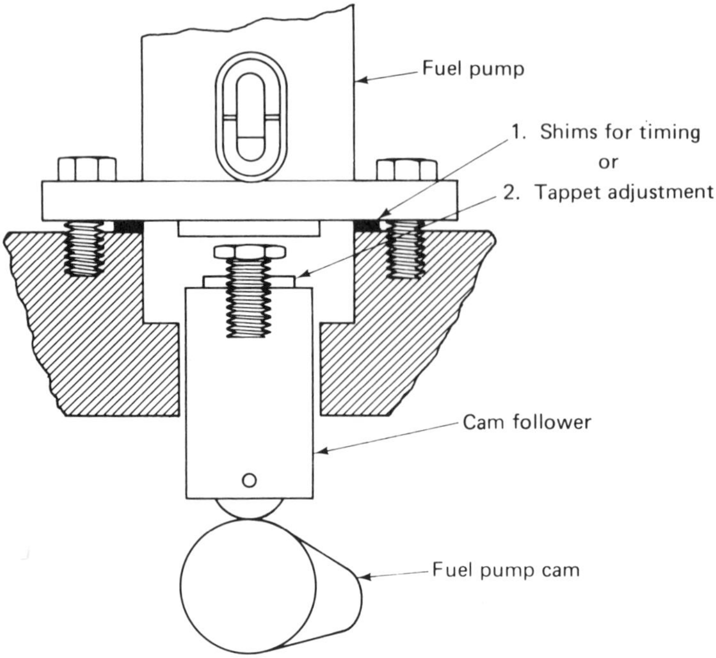

Figure 6-8. Fuel pump adjustments.

6.7 Pump Timing

The fuel injection pump must be timed to the engine. Figure 6-8 shows a fuel pump on an engine with a cam acting on the cam follower. There is a timing window on the pump, (shown in Figure 6-9) with a scribe mark. There is also a scribe mark on the cup that the cam follower acts against in the fuel pump, also shown in Figure 6-9.

To check the engine timing, bar the engine over until just prior to injection; the line on the cup should be near the bottom of the timing window. At the beginning of injection the scribe marks on the cup should be in line with the marks on the timing window. If this is not the case, adjustments should be made either with a tappet adjustment on the cam follower or by the addition or removal of shims between the pump housing and the engine block.

When no timing mark is available, the fuel pump must be timed to the engine. This is accomplished by the free flow method of timing, which is as follows

1. Remove the fuel line and the holding nut.
2. Remove the delivery valve and spring.

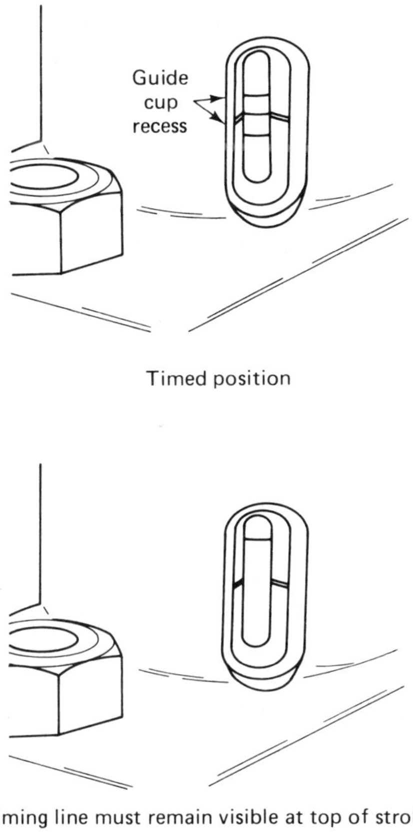

Figure 6-9. Fuel pump timing lines.

3. Replace the delivery valve nut to hold the delivery valve seat in place against the top of pump plunger barrel. Now there is a clear passage for the fuel to flow in and out of the fuel pump.
4. Put the pump in a load position so that the slot is not lined up with the port.
5. Bar the engine over five to ten degrees before injection should occur.
6. Using a hand priming pump, build up the pressure to the fuel pump gradually. The fuel should flow out of the top of the pump. There will be a free flow of fuel, so it will go in through the ports and out through the delivery valve. Bar the engine over a degree at a time until the plunger just covers the port. The free flow of fuel should stop. This is the beginning of injection; thus, we know when the beginning of injection occurs, and the pump is properly timed to the engine.

7. At the completion of the timing procedure replace the delivery valve and spring.

6.8 Injector Nozzles

Let us analyze the fuel injectors in more detail. A metered amount of fuel from the fuel pump is delivered to the inside of the engine cylinder by the injector. The injectors are typically hydraulically operated from the discharge pressure of the fuel pump. Figure 6-10 shows a cross-section of a direct cooled injector nozzle and nozzle holder. The need for nozzles to be cooled will be discussed in a later section.

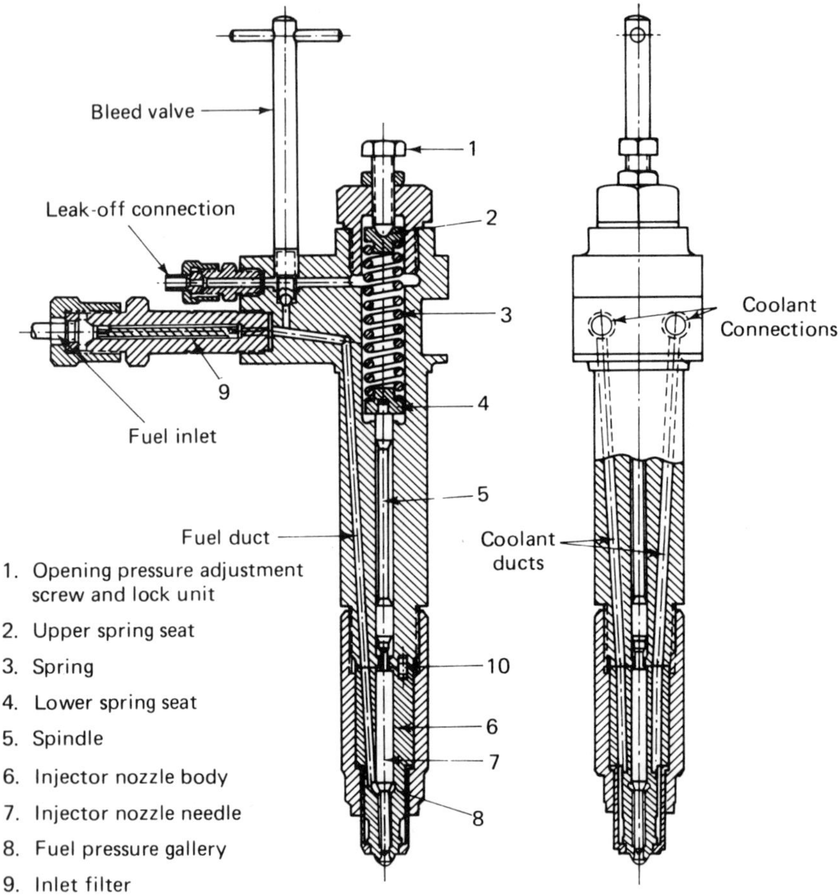

1. Opening pressure adjustment screw and lock unit
2. Upper spring seat
3. Spring
4. Lower spring seat
5. Spindle
6. Injector nozzle body
7. Injector nozzle needle
8. Fuel pressure gallery
9. Inlet filter
10. Dowel

Figure 6-10. Cooled injector nozzle.

Let us follow the flow of fuel as it leaves the metering fuel pump and enters the injector nozzle under high pressure. It first passes through the fuel filter (9) in the nozzle inlet connection. After leaving the filter, the fuel passes through the fuel duct to the pressure gallery (8) near the tip of the nozzle. This is the area below the tapered part of the needle. The fuel pressure is resisted by the spring, which is directly connected through the spindle (5). When sufficient opening pressure for the nozzle is reached, the needle valve (7) snaps off its seat, allowing the flow of fuel through the orifices at the very tip of the nozzle, and injection begins. The reason that the needle valve snaps off its seat is that the fuel, as it first rises, is able to act on a greater surface area, which almost instantaneously overcomes the spring force, pushing the needle valve up quickly. In any event, it is not desirable to have a trickle flow of fuel into the nozzle tip as this can lead to coking.

6.9 Bleed Valve

The bleed valve on the nozzle finds use in two situations. When the injection system has been worked on and there is a possibility of air in the fuel lines, the bleed valve is cracked open and pressure is increased in the fuel system, forcing fuel through the nozzle to the bleed-off connection. In this manner all the air in the lines is removed. All hydraulic systems are based on the incompressibility of fluids. The inclusion of air would result in unpredictable variations in pressure due to the collapse of air bubbles. The second function of the bleed valve is in coordination with the use of engine indicator cards. On large engines indicator cards are taken to determine pressure readings for the cylinder. Two pressure readings are typical: the compression pressure of the engine and the firing pressure of the engine. To determine the compression pressure of the engine while the engine is running, it is necessary to prevent fuel from entering the cylinder. When the bleed valve is opened, there is no pressure build up in the nozzle, so the compression pressure of the engine may be determined. After closing the bleed valve, the engine will again fire and the firing pressure can be determined.

Since the nozzles operate in the hot environment inside the cylinders, there can be a variety of operational problems. Thus, fuel nozzles should be inspected after a set number of hours, the number being provided by the manufacturer. Even if the nozzle is performing satisfactorily, it should be removed for this routine maintenance. Of course, if any engine malfunction occurs, the fuel nozzle is an area that should be checked.

The exact pressure required for nozzle lift-off is determined by the manufacturer of the engine. This is the pressure that will give the best engine performance considering the various factors that go into combustion chamber design.

6.10 Injector Testing

An injector nozzle test stand is used for nozzle inspection so that the nozzle characteristics may be examined. The nozzle test stand allows the injection of clean fuel into the nozzle to determine its performance characteristics. The most frequently occurring problems are dirt, heat, water, and fuel acids reacting with the nozzle, causing the nozzle passages to clog or erode. Just as in an automotive engine, the valve springs may cause problems, but this is rare. The servicing of the nozzle and the injector may be done in three ways: testing, cleaning, or reconditioning. Testing does not mean disassembly; it means testing spray patterns, the opening injection pressure of the nozzle, leakage around the nozzle joints, dribbling from the nozzle tip, and the popping and chattering characteristics of the nozzle. Cleaning is ridding the nozzle of carbon residue and typically involves solvents, wire brushes, lapping compounds, and oils. Reconditioning is not done aboard ship. If a nozzle needs to be reconditioned, it is sent to the factory, where specialized equipment and personnel can undertake the job.

There are five basic nozzle tests. They are: the opening pressure test, the dribble test, the general leakage test, the spray pattern test, and the popping and chattering test. They can be made simultaneously, and all five results should be noted in analyzing the nozzle behavior.

The nozzle will be injecting fuel spray under very high pressure, several thousand psi, and the penetrating power of the spray is sufficient to force oil through human skin, which can cause blood poisoning. The spray from the unit test stand should go into a container, thus preventing oil mist formation in the test room and not allowing a place for hands under the nozzle injection discharge.

The opening pressure test allows the determination of the unseating pressure of the nozzle. The opening pressure is dictated by the engine manufacturer. Typically, the injector manufacturer may have the setting a little high (perhaps by ten percent) to allow for the spring wear. The springs will wear in within a short period of operation, dropping the ten percent overage. New injectors should be tested after some usage to make sure the pressure is correct.

The spray pattern test determines whether the spray pattern from the nozzle tip is distorted as compared to a new nozzle in perfect condition. A blast of fuel from the injector is directed towards a piece of paper, and the pattern is compared to that made by a similar new nozzle. Any distortion in the spray pattern indicates clogged or eroded passages in the nozzle tip. If the nozzle is clogged, then cleaning will remedy the problem; should the problem be caused by an eroded orifice the nozzle body may have to be replaced.

The dribble test indicates poor valve seating, in that the valve is no longer tight and allows a slight amount of fuel to enter the combustion chamber at times other than during the combustion stroke. In this test, the fuel pressure is brought to within 25 to 50 psi of opening pressure and held. The seat should be renewed if the valve allows oil leakage. Of course, general leakage might be observed at any joint between metal surfaces. Should there be any leaks, the surfaces may have to be lapped or renewed.

When the opening pressure is reached, the nozzle valve assembly should move almost instantaneously in a vertical direction because of the pressure differential previously discussed. A distinct popping noise should be heard at this time, indicating that the spring assembly and the spindle are clear of carbon and that the injector opens quickly. Another similar test, using the sound of the valve, is known as the chatter test; a series of sharp pressure pulses to the valve causes it to open and close quickly. The rapid succession of popping sounds is called chattering, and it is a good indication that the valve is not sticking and the nozzle is in good operating condition.

Should an injector fail any of the above tests, it should be disassembled and cleaned. Nozzle parts are typically soaked in an approved solvent and then examined with a magnifying glass for signs of erosion, scoring, or scratching. Clogged orifices are often cleaned with a piece of piano wire whose end has been rounded to prevent scoring of the orifice surface. It is very important that the metal surfaces remain perfectly polished so that there is no leakage. Adequate lubrication is necessary for the proper functioning of the nozzle.

It should be noted that the nozzle tip and the needle valve are a lapped matched pair that cannot be interchanged. If there is a problem with either one, the entire lower unit must be replaced. Lapping is the finishing of working metal surfaces by using an exceedingly fine abrasive material. Any irregularities in two metal surfaces compliment each other and the tolerance or the distance between the surfaces may be less than 0.0001 in. (0.003 mm). Because the clearance is very small there is little leakage and very high pressures may be contained. All high pressure parts are lap fit, metal-to-metal surfaces.

Various nozzle tests have been explained in the preceding paragraphs. From time to time during the testing procedure, nozzles will not perform as outlined in the test procedure. The use of the following troubleshooting guide will help in isolating nozzle malfunction(s). Improperly functioning injector nozzles should be repaired or replaced as they will cause improper combustion, resulting in possible engine unbalance, fuel impingement on piston crown, and other damaging effects.

NOZZLE TROUBLESHOOTING GUIDE

SYMPTOM	CAUSE	REMEDY
NOZZLE OPENS AT TOO HIGH A PRESSURE	pressure screw is incorrectly set	adjust pressure screw
	nozzle valve is seized in the nozzle assembly	replace the complete nozzle
	the spray holes are clogged with carbonized fuel (coked)	clean clogged holes
NOZZLE DRIPS	nozzle valve leaking due to damaged seat or needle valve stuck due to carbonized fuel	if valve seats are damaged, replace complete nozzle; if not, clean with solvent
FUEL SPRAY IS DISTORTED	nozzle tip is carbonized or nozzle valve is damaged	if the nozzle valve is damaged, replace complete nozzle; if carbonization has occurred, clean with an approved solvent
NOZZLE FAILS TO CHATTER	nozzle valve is stuck or nozzle valve seat is leaking	if the nozzle valve is damaged, replace complete nozzle; or clean if carbonized
EXCESSIVE OVERFLOW FROM LEAK-OFF CONNECTION OF NOZZLE HOLDER	too much play in nozzle valve	remove nozzle and clean the surfaces with an approved solvent
	dust, grit, or scale from tubing is between the lap surfaces of the nozzle body and nozzle holder	
NOZZLE TURNS BLUE AFTER SERVICE IN ENGINE	insufficient cooling	check coolant flow through the nozzle, replace nozzle if cooling passages do not allow correct water flow

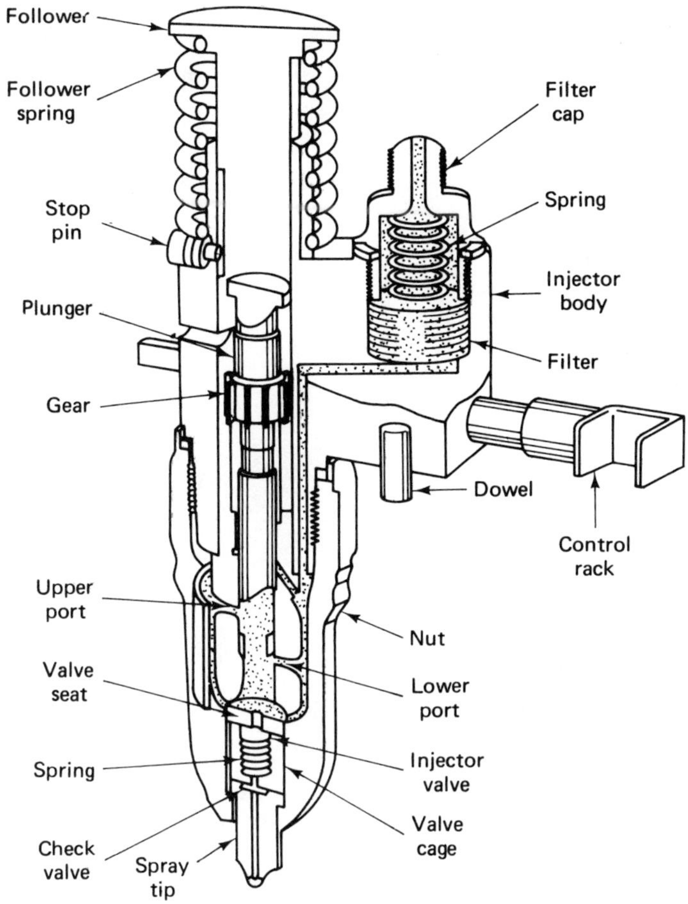

Figure 6-11. Unit injector.

6.11 Unit Injector

The unit injector combines the fuel pump and injection nozzle in one piece, eliminating the need for high pressure fuel lines from the pump to the injector. For safety reasons aboard combatant ships, as well as for space limitation reasons in other applications, the unit injector is widely found on marine diesel engines. It is typically associated with General Motors (Electric Motive Division) engines, although it is found on other engines as well. Typically, the injector is clamped on the center of the head of the cylinder. The clamp is sometimes called the injector crab as it has one leg and two arms and holds the injector in a crablike fashion. There are two fuel lines to the injector; the supply line through which fuel enters the injector body, flowing through the filter and into the body itself; and the return line from the injector which goes back to the supply tank, where the low pressure fuel pump takes suction. The engine has both a supply manifold and a return manifold. The return manifold has a small orifice to maintain constant pressure on the fuel injection system.

Figure 6-11 illustrates a cutaway of a unit injector. The injector is

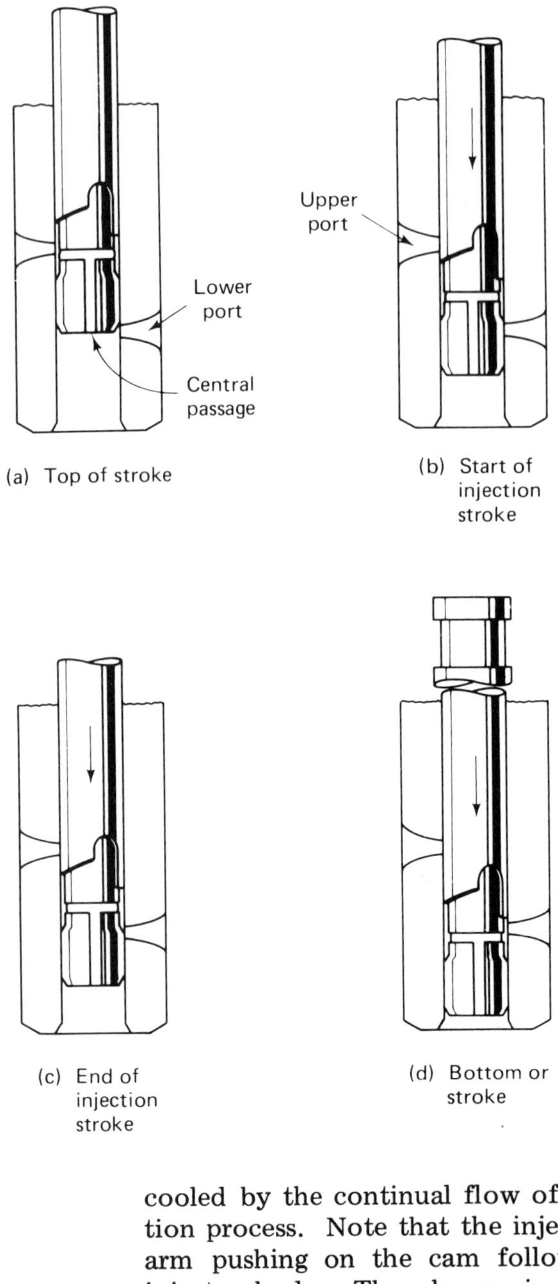

Figure 6-12. Sequence of unit injector operation.

cooled by the continual flow of fuel through it, except during the injection process. Note that the injector is mechanically operated by a rocker arm pushing on the cam follower which activates the plunger in the injector body. The plunger is a rotary plunger pump. The principles of operation of the plunger pump are illustrated in Figure 6-12. There are two ports in the plunger barrel, the upper and the lower port. Within the plunger itself there is a central passage drilled from the bottom of the plunger up to the helix area.

Let us examine the sequence of operation as the plunger travels down the barrel. Motion begins when the rocker arm pushes on the follower, which is resisted by the follower spring, causing the plunger to move downward. At the top of the stroke, there is equal pressure throughout the plunger, as fuel flows in the upper and lower ports and through the central passages. As the plunger moves down, the lower port is covered, as shown in Figure 6-12 (a). When the helix covers the upper port, the pressure is no longer relieved from under the plunger and the start of injection begins. This is illustrated in Figure 6-12 (b). The plunger continues to move downward and the plunger barrel finally uncovers the lower port. This is the end of injection and is shown in Figure 6-12 (c). The plunger continues moving to the bottom of its stroke as indicated in Figure 6-12 (d). The plunger must travel the entire stroke from (a) to (d) as it is positively displaced by the rocker arm. In this case the upper port controls the beginning of injection and the lower port controls the ending of injection. Often the unit injectors are variable beginning and constant ending injectors. There are some, however, that have a slight helix on the lower portion of the barrel causing a variable ending as well. The exact determination of the type of injector is made by engine application. It should be remembered that the barrel and plunger are not interchangeable as the tolerances are very close and they are lap set. If one is scored, the

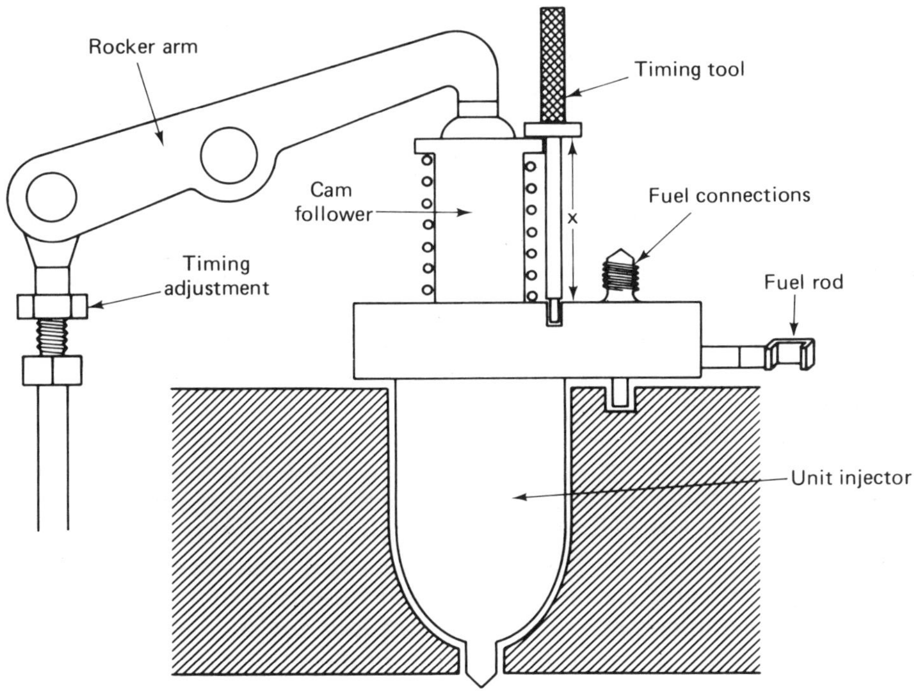

Figure 6-13. GM timing tool.

other will almost necessarily be scored. Even if one is OK, the entire unit must be replaced.

6.12 Unit Injector Timing

The timing of the unit injector is different from that of other types of injector systems. The manufacturer determines where the engine should be barred or jacked over to, setting a given displacement of the injector by the rocker arm. This may be where the exhaust valves are wide open, or it may be a specified number of degrees before top dead center. Figure 6-13 indicates schematically an injector body, a timing tool, and a rocker and pushrod assembly. When the engine is at the specified place, the timing tool is inserted into the injector and the distance *X* should be such that the top of the yoke just touches the top of the follower. If this is not the case, it is necessary to adjust the pushrods slightly to correct the distance to the specified amount.

6.13 Unit Injector Testing

The testing of the unit injector is somewhat different from the testing of other injectors because of the construction of the injector. The tests themselves are the same. There is a separate test stand for unit injectors. With the injector in the test stand, it is necessary to build up the pressure on the injector body to find out at what pressure the injector opens. For a General Motors unit injector it should be between 450 and 750 psi. If the injector opens below 450 psi, it is necessary to renew the tip assembly. There is no mechanism for controlling the opening pressure as it is set by the manufacturer in the design of the spray tip. This is the opening pressure test. The next test determines the spray tip characteristics. This test is slightly different from previous spray pattern tests, in that the fuel rack must be actuated so the injectors are in the load position. The injector plunger is actuated rapidly to cause the fuel to spray out from the tip, and the spray pattern is determined for the orifice. The difference is that in this case the control rack must be actuated so that there will be a build up of pressure within the injector to open the injector tip. The test is the same as for other injectors. If there is leaking, the only recourse is to replace the tip. It is no longer considered acceptable practice to lap the various valves and valve seats. When performing the previous tests, the body of the injector should be examined for leaks. If there are leaks at gasketed joints, the gasket should be replaced; if leaks occur in lap fit areas, the injector body should be replaced. Replacement is easily accomplished as the injectors are widely used and the old injector is accepted as a trade-in for a new or reconditioned injector.

7

Governors

This chapter will discuss the mechanism by which the system and the engine are governed so that the engine will meet its load demands by proper fuel metering. Speed and load regulation are accomplished through a governor.

A governor is a speed sensitive device used to control or limit the engine speed. There are two types of governors: direct acting, or mechanical, governors; and indirect, called relay or hydraulic governors. In analyzing these governors, principles of operation and certain troubleshooting procedures will be emphasized. Before that point is reached, however, it is important to have a general understanding of how a governor operates and of the differences between the mechanical and hydraulic governors.

7.1 Classification by Function

Governors may be further classified by function, of which there are three. The first function is regulation. There are two categories: constant speed governors, typically used on diesel generator sets where it is important to maintain a constant engine speed regardless of load, and variable speed regulating governors, which maintain a set output speed at any given load, although that speed may be varied with load through the use of the governor. These are typically used on main propulsion engines. Thus, the governor will maintain constant shaft speed at half ahead as well as at full ahead regardless of load changes on the propeller. Figure 7-1 illustrates the speed-time diagram for regulating governors.

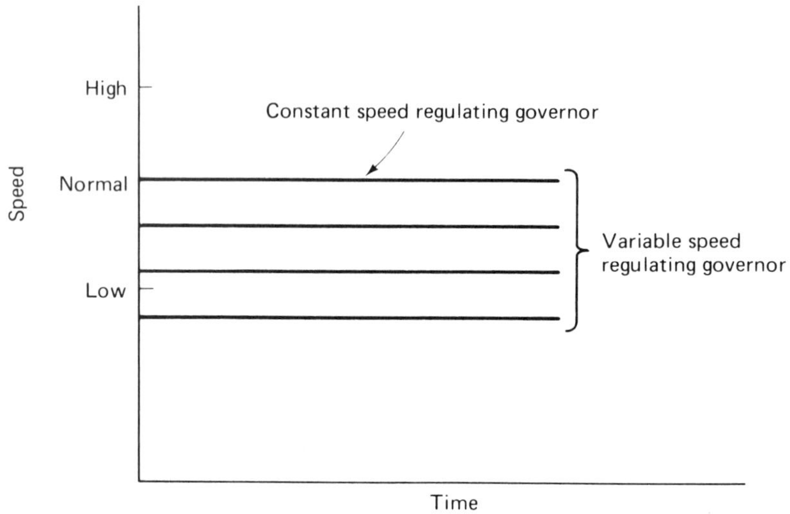

Figure 7-1. Speed-time diagram for constant and variable speed governors.

The second general function is to limit speed. A speed limiting governor will regulate both maximum speed and idling speed; all other speeds are manually set. This type of governor is found aboard small vessels and is similar to the governors found in truck and automotive engines. Its purpose is to prevent the engine from overspeeding when load is suddenly taken off the engine, and from stalling when the throttle is in the lowest possible position. Aboard ship, the throttle position is typically set by hand; in automotive and truck service the foot on the fuel pedal establishes engine speed. The chief advantage of the speed limiting governor is its simple design; it is less expensive to design and install than hydraulic governors. Figure 7-2 illustrates a speed-time diagram for this class of governor. In the example shown, the speed limiting

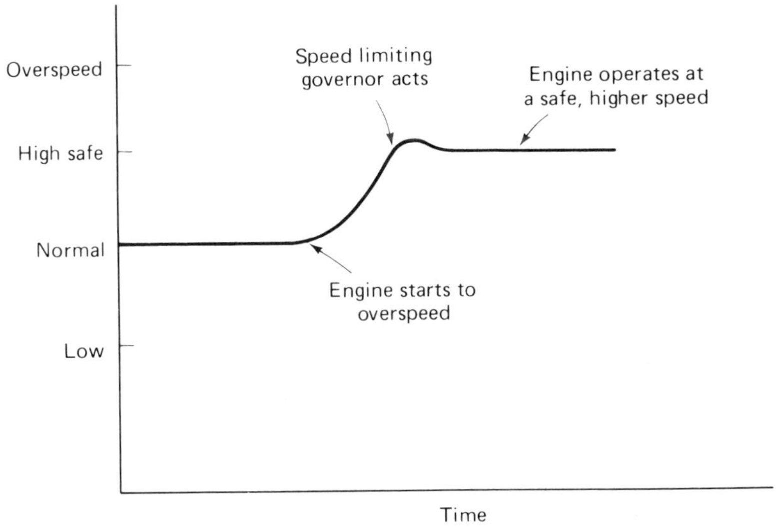

Figure 7-2. Speed-time diagram for a speed-limiting governor.

governor allows the engine to operate at a higher, but still safe, speed. Usually an alarm will notify the operator of any control malfunction.

The third general function is engine overspeed protection; there are two types of governors serving this function, and either may be used in conjunction with speed regulating governors to provide protection in case the latter should fail. The two types of overspeed trip governors are manual and automatic reset. Figure 7-3 (a) illustrates a speed-time dia-

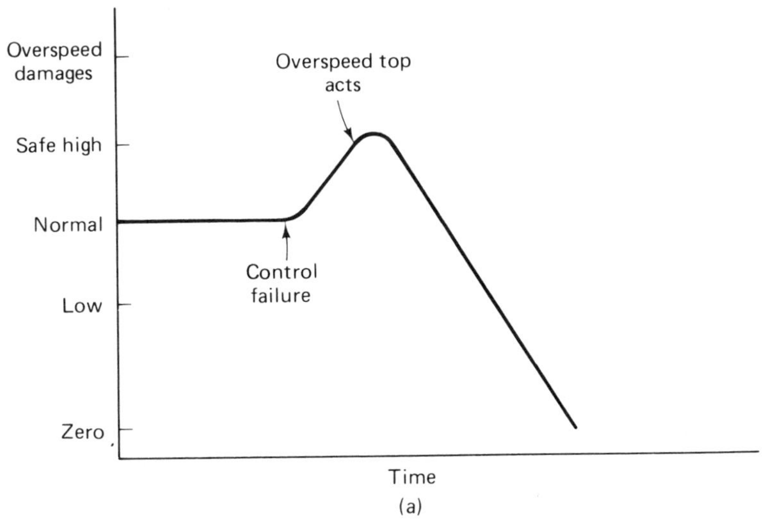

Figure 7-3. (a) Speed-time diagram for an overspeed trip governor.

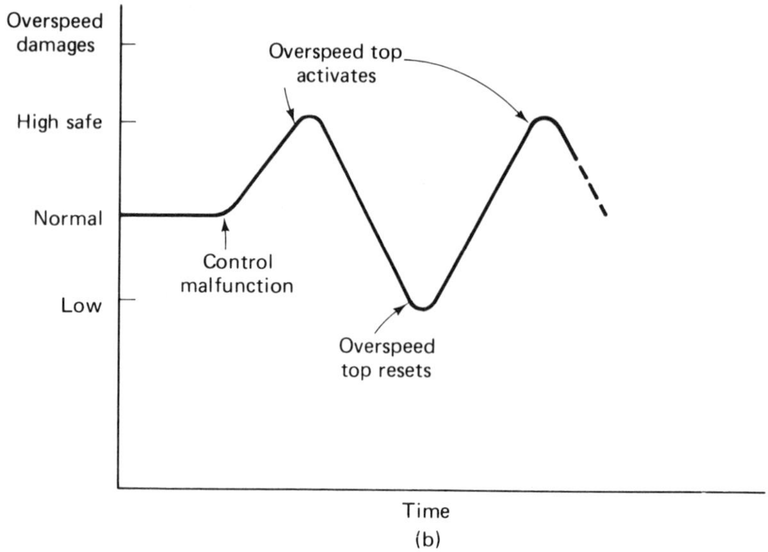

Figure 7-3b. Speed-time diagram for automatic reset/overspeed trip governor.

gram for the manual type unit and Figure 7-3 (b) shows an automatic type. The manual reset type will stop the engine if overspeed occurs, and the operator must reset the trip to start the engine. The automatic type does not stop the engine, and the governor resets itself. Either type, however, will sound an alarm, indicating that a malfunction has occurred. Both types act by stopping the flow of fuel to the engine. There is no regulation of fuel supply. Both types may be found on engines with all types of speed control governors as a safety device for overspeed protection; which one is installed depends on the desires of the manufacturer and/or operator.

Before proceeding further with the discussion of governors and control systems, a few definitions of some of the more important terms describing governor and engine operation are in order.

7.2 Definitions

Speed droop: A progressive drop in speed as load is picked up by the prime mover from no load to full load without manually changing the speed setting. It may be expressed in rpm or in percentage. Speed droop in rpm is equal to no load rpm minus full load rpm. Speed droop in percent is equal to no load rpm minus full load rpm divided by full load rpm times 100.

$$\%\ \text{Speed Droop} = \frac{\text{No load rpm} - \text{full load rpm}}{\text{Full load rpm}} \times 100$$

Sensitivity: The smallest speed change that will induce the governor to alter the flow of the energy medium. The term energy medium, rather than fuel, is used because governors are used on all types of devices, involving many energy sources.

Stability: The ability of the governor to correct a speed disturbance with a minimum of false motions.

Compensation: A mechanical and/or hydraulic action that prevents overcorrection of the energy medium (fuel) supply. Compensation in a governor produces transient speed droop, a momentary speed droop during a speed correction.

Dead band: A narrow band of speed variation through which the governor makes no correction of the energy medium (fuel). This is common in heavy mechanical governors and refers to the sensitivity of the governor.

Hunting: A rhythmic variation of speed that can be eliminated by blocking the energy medium (fuel) supply manually or with a load limit control, but which reappears when the engine returns to governor control.

Isochronous: Constant speed, specifically the same average speed regardless of load.

Overspeed: A speed greater than the selected normal speed or greater than the highest speed attained for a specified load selection.

Parallel operation: Two or more prime movers (diesels) are electrically interconnected and driving a common shaft and load.

Smoke limiters, load limiters, maximum fuel stops: Devices used to limit the maximum amount of fuel delivered to the engine.

Maximum speed stop: A device used to limit the spring force on a regulating-type governor.

7.3 Mechanical Governors

The mechanical governor is linked directly to the fuel pump racks so that the action of the flyweights on the governor will cause the fuel pump rack to move, changing the amount of fuel that enters the engine. A schematic of a mechanical governor is shown in Figure 7-4. The governor is connected directly to the engine by a small driveshaft, which may be geared up to drive the flyweight at a speed greater than engine speed. Such an arrangement is common in high and medium speed engines. The flyweights are attached to the ballhead, which meshes with the driveshaft gear. The flyweight's outward movement, due to centrifugal force, is resisted by a speeder spring. The spring moves, and this change of posi-

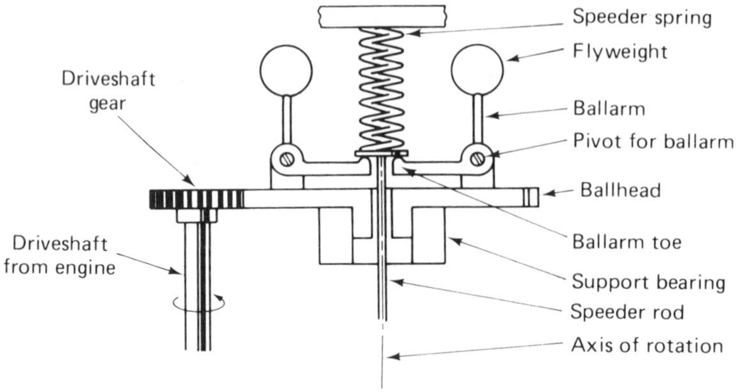

Figure 7-4. Mechanical governor.

tion causes the speeder rod to move. The speeder rod is directly connected to the fuel pump. In the sketch shown, as the rod moves down, the fuel flow to the engine increases.

Hydraulic governor flyweights also sense the engine speed. In their case, however, they actuate a pilot valve, and hydraulic oil pressure causes the fuel pump racks to move. Thus the flyweights indirectly affect the fuel pump rack position.

Let us analyze mechanical governors in more detail. Figures 7-5 (a), (b), and (c) illustrate a mechanical governor at various engine speed load configurations. What must a mechanical governor consist of to operate? First, it must have a set of flyweights driven by the engine, typically at a higher rpm than engine speed to increase their sensitivity to small changes in engine speed; second, there must be a governor spring opposing the centrifugal force of the flyweights, allowing the governor to reach an equilibrium value; and third, the flyweights must actuate, through a direct linkage, the fuel pump rack which, in turn, meters the fuel to the injectors. These components are shown schematically in Figure 7-5. In this case, a pipe is used in lieu of a fuel rack, with the positioning of the fuel rack simulated by an opening or closing of a valve in the pipe, permitting more or less fuel flow. Figure 7-5 (a) shows the equilibrium value of the mechanical governor at fixed load conditions. Assume that the engine is running at half load. Note that the flyweights are in the vertical position, with an equilibrium or static force balance existing between the governor spring and the flyweights. Assume now that the engine goes from half load to full load and in doing so slows down; the centrifugal force on the flyweights decreases. The result will be as shown in Figure 7-5 (b). The valve opens further, allowing an increased flow of fuel, so the engine can run at the higher load. Note, however, that even when a new equilibrium is achieved, the flyweights

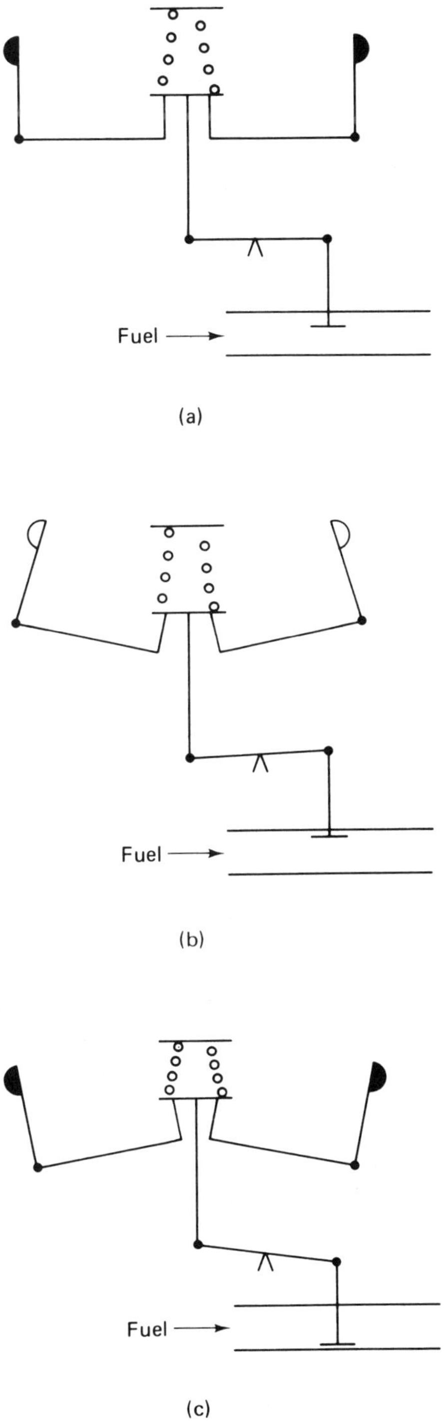

Figure 7-5. Flyweight variation with load for a mechanical governor. (a) Equilibrium position; (b) increased load, increased fuel flow; (c) decreased load, decreased fuel flow.

will not be able to go back to the original vertical position they had at half load. This is because the valve must be open further to allow the greater fuel flow for the higher load. Thus, the mechanical governor automatically has a speed droop, a decrease in speed from no load to full load. Now assume just the opposite, that the engine goes from half load to no load. The no load position is illustrated in Figure 7-5 (c). Notice that the engine rpm increases, causing the flyweights to move further out, and shifting the valve to close off the fuel supply to the engine. The engine will be running at a higher rpm at the no load condition than at the half load condition.

The flyweights must be large enough and/or the spring of the mechanical governor must be strong enough to cause the fuel pump rack to move. These requirements are adequate on small engines where the parts to be moved are also small and the force required by the spring or flyweights is not large. The inertia of all parts must be small to adapt to the inertia of the flyweights. As the flyweights become larger and heavier, they are not so sensitive to changes in engine speed because of their inertia. However, they do create a large enough force to move larger fuel racks, which slight and more sensitive flyweights could not do. This is particularly so on large engines, where the fuel pumps and racks may weigh 75 pounds or more and can be moved only by great force. The engine would fluctuate in a larger dead band than is desirable, causing hunting. The mechanical governor has speed droop and cannot be used for situations where isochronous control is desired.

7.4 Hydraulic Governor

The hydraulic governor, shown in Figure 7-6, consists of three main parts: a speed-sensitive section, which senses the engine speed and tries to maintain a constant engine speed; a power section, which actuates the fuel racks controlling the flow of fuel to the engine; and a compensating section, which compensates for the difference in reaction time between the engine and the governor. The governor tends to react very quickly because it is hydraulic, has few parts, and the inertia of those parts is low. The mass and inertia of the engine, on the other hand, are quite high; it will not react so quickly to speed changes. There must be a circuit or mechanism built into the governor to compensate for the differences in reaction time of the engine and governor so they will act in accord, rather than at odds, with each other.

7.5 Engine Time Lag

Engine time lag is the time from the moment the load and/or speed changes of the engine occur until the engine is again running at the correct

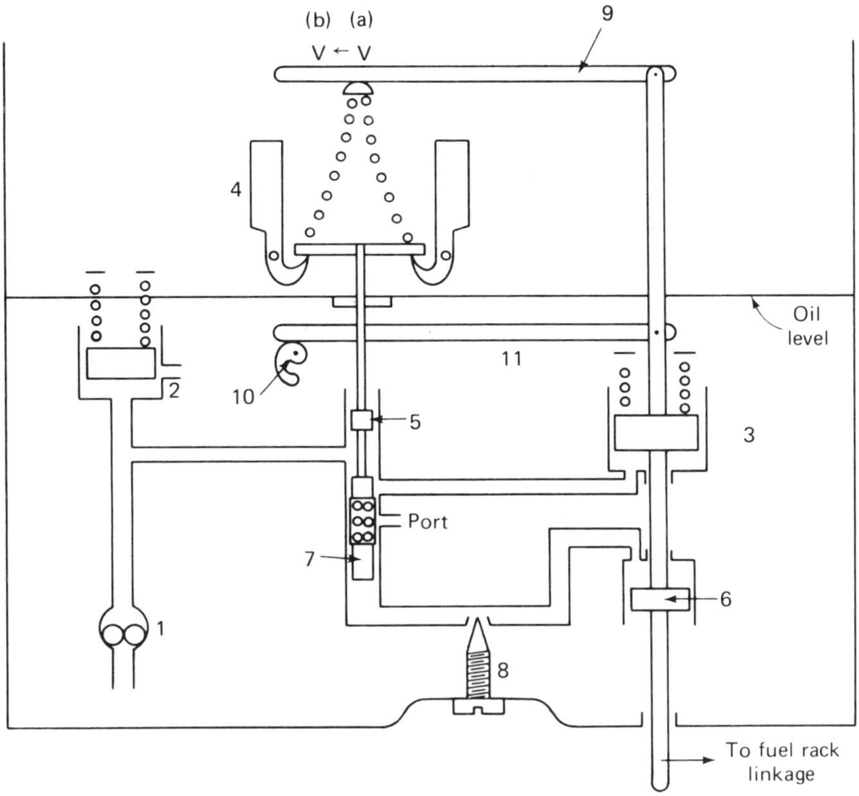

Figure 7-6. Schematic of a hydraulic governor.

load or speed. Various factors affect the time lag of the engine. There are five factors:

1. The delay of the governor in sensing the change in engine load/speed (manifested by the small flyweights of the governor changing velocity).
2. A delay, within the mechanism itself, from the moment the flyweights sense the action to the moment the governor moves to react. This is caused by the inertia of the governor's parts.
3. The time from when the governor causes the fuel pump settings to be changed until the fuel pump is changed.
4. The time that it takes the corrected fuel charge, once the pump is reset, to enter the engine cylinder. In this case, fuel will be admitted at the old setting, and while this is occurring correction is being made with rotary plunger pumps. There will be some delay before the correct amount of fuel is admitted for the new load/speed conditions.

5. The time required for the correct fuel charge to be converted into the correct engine load/speed. In this instance the fuel will be correct, but it will take time for the power to be transmitted from this new amount of chemical energy into new shaft energy or mechanical energy in the engine.

7.6 Hydraulic Governor Operation

Let us consider Figure 7-6 in more detail and go through the numbered items, as indicated in the figure. Item 1 is the gear pump. The gear pump and the flyweights are directly driven from the engine gear train. These may be at a higher or lower rpm than the engine. The gear pump takes the oil supply from the sump and delivers it at a higher pressure to item 2, the accumulator. The accumulator stores oil under pressure. Its pressure is constant, fixed by the spring tension, which is determined by the manufacturer. The gear pump is a positive displacement pump and will increase the oil pressure in the cylinder above the set value. As the accumulator piston rises, oil leaks out through the port in the accumulator to the oil sump of the governor. Item 3 is the power piston, which activates the fuel pump rack. The spring on top tends to push the engine fuel rack to the closed position, while an increase in oil pressure on the bottom tends to cause the fuel rack to supply more fuel to the engine, increasing engine speed. Items 1, 2, and 3 make up the power section of the hydraulic governor. Items 4 and 5 make up the speed-sensitive element of the governor. Item 4 is the flyweight and speed-sensitive spring, and item 5 is the pilot valve and its ported bushing. The pilot valve is directly attached to the flyweights through the speeder rod. Items 6, 7, and 8 are the compensating section of the governor. Item 6 is the compensating transmitting piston, item 7 the compensating receiving piston, and item 8 the needle valve. The needle valve controls the flow of oil in and out of the closed hydraulic compensating system.

Let us analyze an isochronous hydraulic governor of Figure 7-6, which has a built-in compensating system. There are, however, hydraulic governors which are not isochronous. In that case it is not necessary to have a compensating system, as the speed droop compensates for the variation in engine speed. We should also note that this governor is not a specific manufacturer's; it is composed of typical generic parts found in all hydraulic governors, but it does not indicate the complexity of the actual linkages and mechanisms. Its purpose is to better explain how the governor operates. In practice, the repairs made by operators on a hydraulic governor are almost zero. If a hydraulic governor fails, the accepted practice is to send it back to the manufacturer for repair. Two items that are on the diagram but are not part of the three sections mentioned are item 9, the speed droop adjustment arm, and item 10, the load limit cam. In this particular case the speed droop arm has a movable fulcrum so that when the fulcrum is in position (a) the governor acts as an isochronous governor having zero speed droop. When the fulcrum is moved to position (b) there is speed droop, which may be set from zero to some maximum value, perhaps 5%, indicated by point (b). This is a manual adjustment; the control is found on the outside of the governor. The load limit cam may be rotated to limit fuel supply to the engine. In the position shown, the engine can run to any load from zero to full. If the cam is rotated in a clockwise direction, so that the load limit arm moves up, it will prevent the flyweight from moving the pilot valve, thus preventing the power piston movement to the maximum fuel position. It is the movement of the pilot valve under load changes that enables the governor to work. With these details in mind, let us analyze changes of load and the reactions of the governor.

7.7 Compensation

Let us assume that the engine is running at half speed at a fixed rpm. In this case the only way an equilibrium position can be maintained is for the pilot valve (5) to cover the port to the power piston. The flyweights must be in a vertical position for equilibrium to be achieved. Assume that there is an increase in engine load; the engine rpm decreases, and the fuel rack linkage moves up in the governor, allowing more fuel to flow to the engine. Since the engine has decreased in speed, the flyweights (4) will collapse, allowing the spring pressure to push down on the pilot valve, uncovering the port. The oil passage from the gear pump opens, allowing a flow of oil and an increase in oil pressure to the power piston, causing it to rise. As the power piston moves up, the compensating transmitting piston (6) also moves up, raising the pressure of the oil in the compensating system. This increase in oil pressure acts on the compensating receiving piston (7), causing it to push the spring, which in turn is attached to the pilot valve. This force pushes the pilot valve toward the equilibrium position, closing the port from the power section. Simultaneously, there is a leakage of oil from the compensating system through the needle valve back to the sump. As the engine speed reestablishes itself, the equilibrium position of the flyweights returns. The flyweights are vertical, and the pilot valve moves down, closing off the port from the power section. In this case, the engine is running at a new and higher load; however, the engine speed remains constant. The port will be closed and equilibrium reestablished when the flyweights are in the vertical position. The adjustment of the needle valve is critical to the compensating system. The adjustment must be fixed such that the leakage through the needle valve is coincident with the resumption of speed by the engine. This adjustment is initially made by the operator when the governor is installed and typically another adjustment is not necessary. The most common exception would be if dirt gets in the orifice of the needle valve.

Let us now assume that the engine is operating at half load and that a change of load occurs; however, in this case the engine load is removed. Intuition tells us that the fuel rack leakage must move down, limiting the fuel supply to the engine. Let us observe what happens. As the load is removed, the engine will overspeed, causing the flyweights (4) to move out. This will cause the pilot valve to move upward, allowing a discharge of oil from the power piston through the port connecting it and the compensating receiving piston. Because the pressure under the power piston is lower, the spring force acts to push the fuel rack down, closing off the fuel supply to the engine. The compensating piston retards this motion as the oil pressure in the compensating circuit is now below that of the surrounding sump. Oil flows in through the needle valve to the compensating circuit. The compensating receiving piston (7) tends to pull the pilot valve down, covering the port to reduce the flow of oil from the power piston. As previously, the only way for equilibrium to be established at the new no load position is for the flyweights to be in a vertical position. The compensating circuit always acts to reduce the effect of the power section. This balances the differences in reaction time between the governor and engine. The needle valve must be set for the peculiarities of each engine to balance these differences.

Notice the small metal tab attached to the speeder rod (11). Should the load limit cam be rotated to its maximum height, pushing the load limit arm to its most vertical position, this tab will prevent the pilot valve from moving down. This prevents a flow of fuel to the engine, thus causing it to stop. This occurs because the spring force in the power piston cannot be resisted by oil pressure and the fuel rack linkage is in the closed position.

7.8 Speed Droop Adjustment

Consider what happens when the speed droop adjustment arm is moved so that the fulcrum is at point (b) rather than at point (a). The speed droop adjustment changes the spring force that the centrifugal force must overcome to reach the vertical position. The flyweights must be in the vertical position for the pilot valve to be closed. At no load the spring force will be greater because the arm (9) is in a lower position, slightly compressing the spring. Thus, the centrifugal force on the flyweights must be greater. Conversely, at full load rpm with the arm (9) in a higher position, the spring force is less, and less centrifugal force is needed to move the flyweights to the vertical position. Thus, the force is not uniform over the speed range of the engine. There is a variation of rpm from full load to no load. This variation is defined as speed droop. In comparing the hydraulic governor and the mechanical governor, note that the hydraulic governor is more complex; however, it may be used as an isochronous governor, and it allows the speed-sensing section to be very sensitive and independent of the power section. Thus, it can be used on large engines where a great deal of force is necessary to overcome the inertia of the fuel pump and its components.

7.9 Flyweight Springs

An aspect of the mechanical and hydraulic governors that has not been discussed is the shape of the spring resisting the centrifugal force of the flyweights. Typically, this spring is not cylindrical, but conical, and it is sometimes referred to as a trumpet-shaped spring. Figure 7-7 is a diagram of load change versus deflection for cylindrical and conical springs. This figure shows the percentage change in load the engine experiences, e.g., from half to full power. It is desirable that the governor be sensitive to small load changes, to prevent having a large dead band. A cylindrical spring has a linear deflection and the deflection change for a small load change is not large. The spring curve of a conical spring is nonlinear, allowing large deflections for small load changes. For large load changes, the deflection change need not be proportionately large. The nonlinearity of the conical spring reduces its total deflection for large load changes, which is not the case for cylindrical springs. This allows the governor to have greater stability at all loads and reduces the dead band for small load changes.

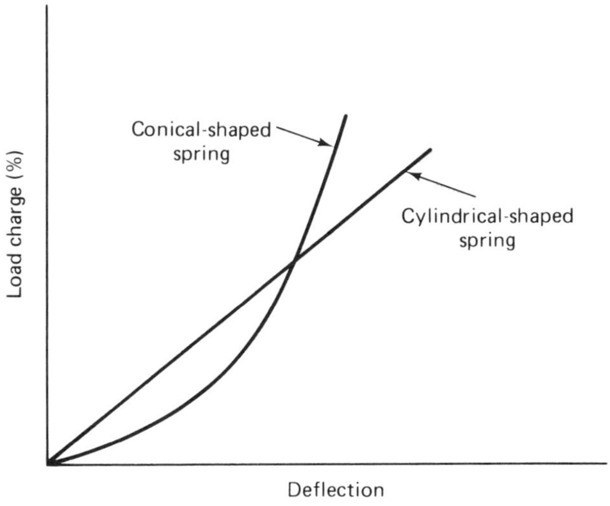

Figure 7-7. Effect on spring shape on load-deflection curve.

7.10 Speed Droop

On diesel-driven electrical generators, the engine must have speed droop to allow the generators to share the load. Consider the isochronous governor discussed earlier. In that case the engine runs at constant speed for all values of load. There is no way to know the engine load as a function of speed. The engine could be under 25% or 100% load and the difference could not be observed by knowing the engine speed. As an observer cannot see the difference in load, another engine cannot sense the difference to allow sharing the load. This is why speed droop is necessary. Figure 7-8 illustrates a speed droop curve of 3.4% for two diesel engines, *A* and *B*. Consider two similar engines driving two similar generators; assume that engines *A* and *B* are at 50% load. We wish to change the load of the system and observe engines *A* and *B* to see how the load and rpm have shifted. As more load is added to the system, we will assume that either engine can take all this additional load momentarily. Let us assume that engine *A* does so. Engine *A* will slow down, engine *B* is momentarily running faster, and the faster machine will take the load from the slower machine; thus, the machines will stabilize at a new common rpm. The electrical reason for the faster moving machine's taking the load is that both machines are connected to a common electrical connection between the two generators known as a bus bar. As engine *A* slows down, the cur-

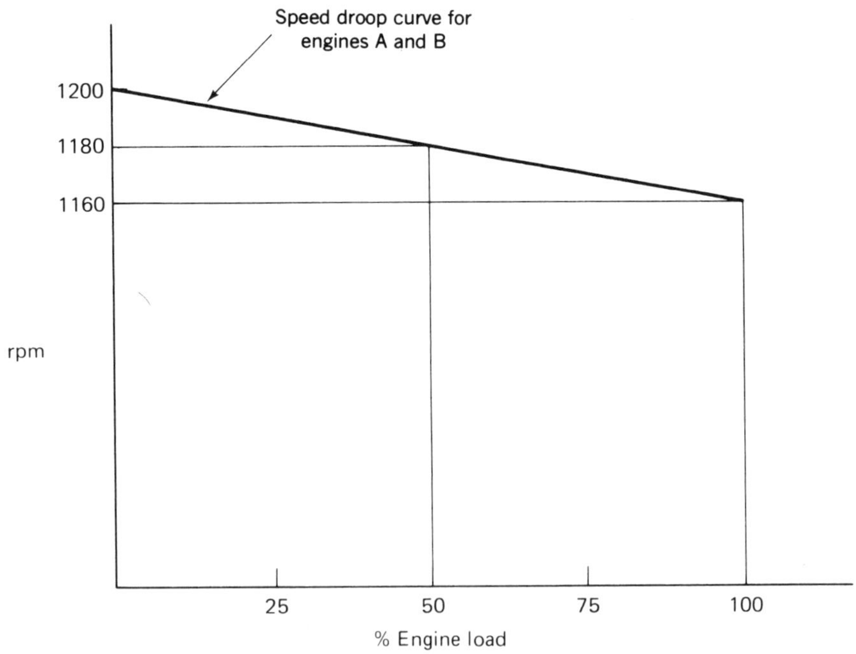

Figure 7-8. Speed droop curve.

rent in the common electrical bus tends to increase; however, this current also affects engine *B*. It senses a greater load and slows down because of the greater current, or load.

There may be times when it is desirable to have engines paralleled while having different speed droops. This situation is illustrated in Figure 7-9, where engine *A* has a 40 rpm, or 3.4%, speed droop and engine *B* has an 80 rpm, or 7.1%, speed droop. Consider the case where engine *A* is at 100% load, which sets the system speed, as the generators must run at the same speed. Engine *B* will be at 1160 rpm because of this synchronization. With the speed droops selected, engine *B* would be at 50% load. Similarly, as the load decreased, engine *A* would be at 50% load and engine *B* would be at 25% load. Thus, we can regulate the load division between engines by changing the speed droop of the engine. Typically, the speed droops should be as near to one another as possible; however, there could be an instance where engine *B* is nearing repair or overhaul and should be running at less than full load. Thus, the speed droop can be adjusted so that engine *B* would take between 75% and 90% of the load when engine *A* is at 100%. The 50% loading used in the previous example is not indicative of the speed droop setting that would be used in practice.

Figure 7-10 illustrates a situation where one machine operates with an isochronous governor and the other machine operates with 40 rpm speed droop. In this case the system will be running at the frequency, and

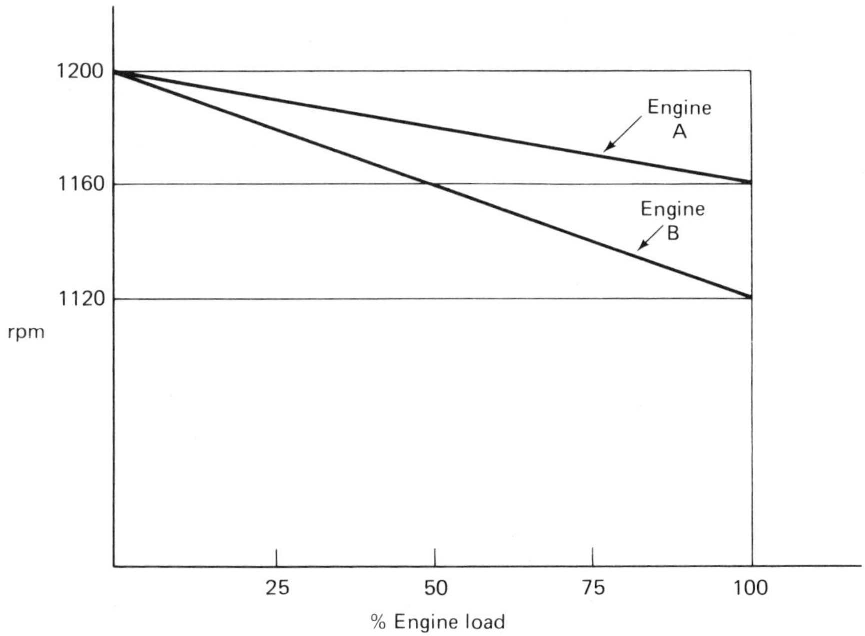

Figure 7-9. Effect of speed droop on engine load distribution.

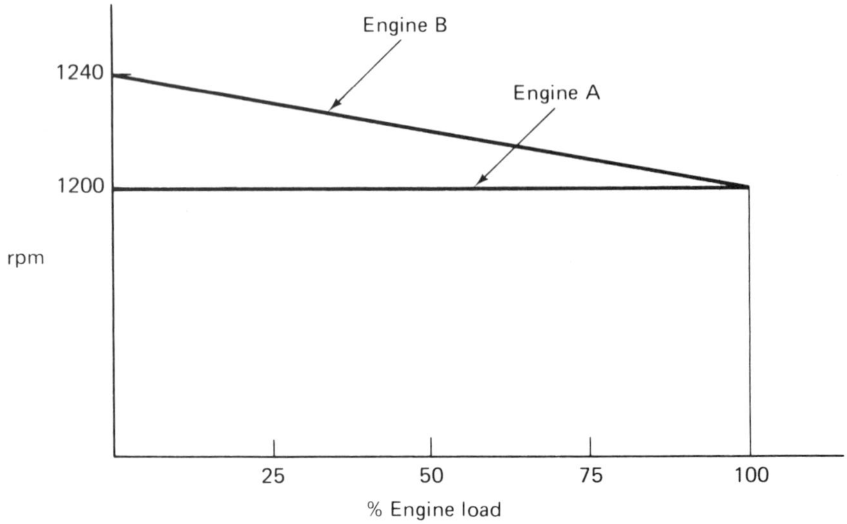

Figure 7-10. Effect of parallel operation for engines with and without speed droop.

hence, the rpm, of the isochronous machine. The system speed must be 1200 rpm, because this is the speed at which engine *A* will run. If this is so, engine *B* will be fully loaded and engine *A* will assume the variable system loads, such as pump, fans, and cargo gear starting and stopping. It is important that the system be at more than 100% of one engine; if this were not the case, it would be possible for engine *A* to be tripped off the line electrically because of reverse power from engine *B*. In this situation, engine *B* senses that engine *A* has no load and will try to drive it as a motor. There is a reverse power relay on all generators that prevents them from being motorized. This would trip diesel generator *A* off the line. Should this occur, the load subsequently becoming greater than the maximum load required by engine *B*, engine *B* will trip off the line because of overload protection.

What about cases with significant load changes, say from 75% to 125% of the load of one engine, where two diesel generators are needed, but the situation illustrated in Figure 7-10 cannot be used? Figure 7-11 shows a method by which we can satisfactorily comply with the system loading. In this case we must run two generators. Generator *A* will be the isochronous generator, allowing the system to operate at the constant frequency demanded of the electrical system, but the speed droop of engine *B* has been changed, allowing it to run only from 0% to 50% of its maximum load. It can go down to only 1200 rpm. How can this be accomplished? In Figure 7-10, engine *B* had a no load rpm of 1240. In

Figure 7-11 engine *B*'s no load speed has been reduced to 1220 rpm. This is accomplished from the switchboard by decreasing the frequency control switch, which lowers the no load speed of the diesel generator. While it is possible to go further, creating different situations where different speed droop considerations must be implemented, let us stop at this point. The important thing to remember is that for best fuel consumption an engine should run at full load or as near the maximum load as possible. Engines are designed to have their optimum specific fuel consumption under full load conditions; thus, to minimize fuel consumption, the engine should be fully loaded or taken off the line. Typically, one engine might be fully loaded and the other partially loaded, accounting for variable loads.

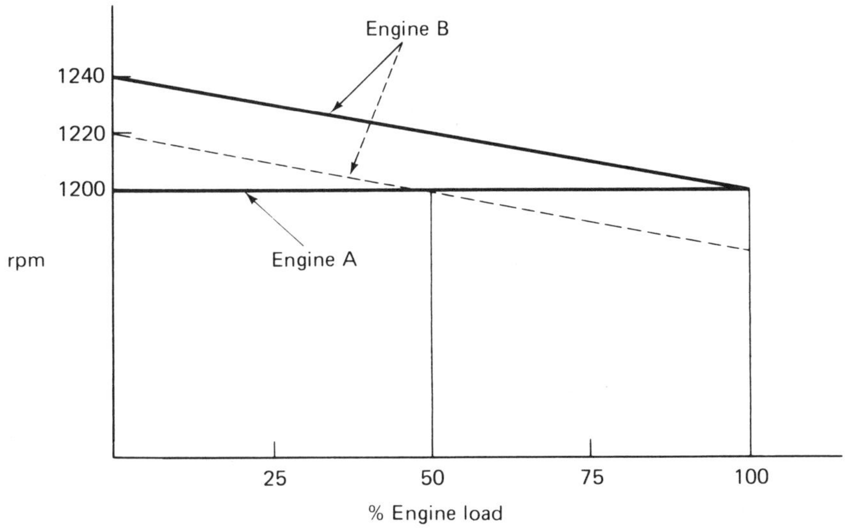

Figure 7-11. Speed droop diagram.

7.11 Variable Speed Droop Mechanisms

Variable speed droop is an important quality in both isochronous and mechanical governors. We have discussed variable speed droop in the hydraulic governor. It is also possible in a mechanical governor. Figure 7-12 is a schematic sketch showing how a mechanical governor may have a variable speed droop. The flyweights (2) act on the collar (1), which in turn rotates the lever (6). As the flyweights increase in speed, the engine speeds up. The flyweights move up, and the lever (6) moves from the maximum fuel position to the minimum or no fuel position. The movement of the flyweights is resisted by the spring (3); thus, as the flyweights go out, the spring is stretched, and the governor exhibits the speed droop

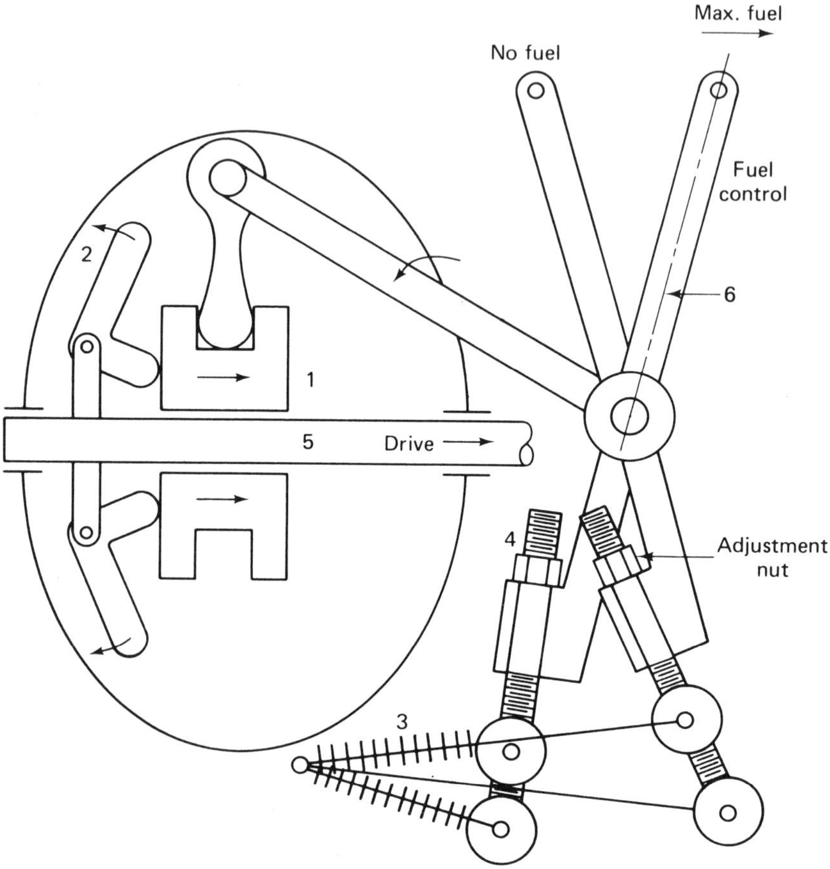

Figure 7.12. Mechanism to allow variable speed droop for mechanical governors.

typical of mechanical governors. The adjusting nut (4) may be turned to increase the spring tension that the flyweights act against. This gives the variable speed droop required.

The no load setting, i.e., speed droop adjustment, of the hydraulic governor may be examined in a little more detail. The schematic diagram did not illustrate the mechanism for adjusting the fulcrum. Figure 7-13 details the fulcrum area in a schematic fashion for a hydraulic governor. Typically, item (5) is a knob located on the device or on the generator control panel; it is called the frequency control device. Instead of there being a knob, the control may be driven by an electric motor. Be it a knob or an electric motor, it will move the collar (6), which in turn changes the force that the fulcrum will exert on the cylinder (4). The cylinder (4) is sometimes known as a speeder plug. By increasing or

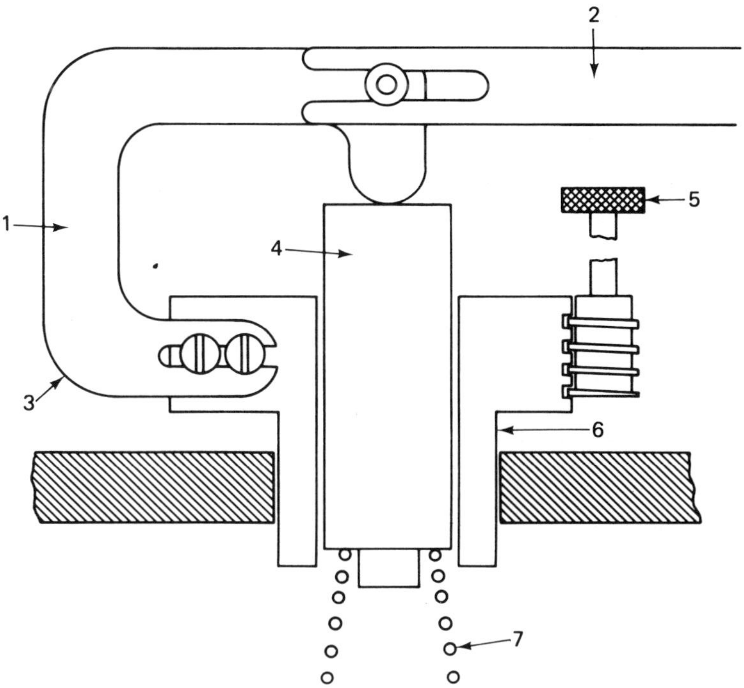

1. Speed droop adjustment
2. Speed droop arm
3. Speed droop fulcrum adjustment
4. Speeder plug
5. Speed adjusting knob and gear
6. Speed adjusting movable sleeve
7. Governor speeder spring

Figure 7-13. Schematic of hydraulic governor speed droop adjustment.

decreasing the force of the fulcrum against the speeder plug, the no load speed will be changed. The greater the force, the greater the no load speed, and vice versa. The hydraulic governor only works by the fly-weights coming to a vertical position; thus an increase in force will require a greater rpm to establish a vertical position. Thus, the no load speed of the governor and its speed droop has been raised.

8

Cooling and Lubrication Systems

8.1 Cooling Systems

The cooling systems of diesel engines have two main functions. The first is to remove between 30 and 35% of the heat from the engine; the second is to cool the engine's lube oil. The lube oil cooling may be accomplished by either fresh or salt water.

Open System

There are two general types of cooling systems. One is the open system, in which sea water is in direct contact with the engine. It is not often found aboard diesel engined vessels anymore because of problems associated with it. The maximum outlet temperature is limited to about 120°F since, if the temperature exceeds that, salt will precipitate on the engine block. Furthermore, cold sea water can create thermal stresses within the engine.

Closed System

To overcome the disadvantages of the open system, a closed cooling system, using fresh water, was developed. Figure 8-1 is a schematic diagram of a closed cooling water system. The fresh water is circulated through the engine starting at the bottom, where the engine is coolest and the water is also coolest. It circulates vertically to the top

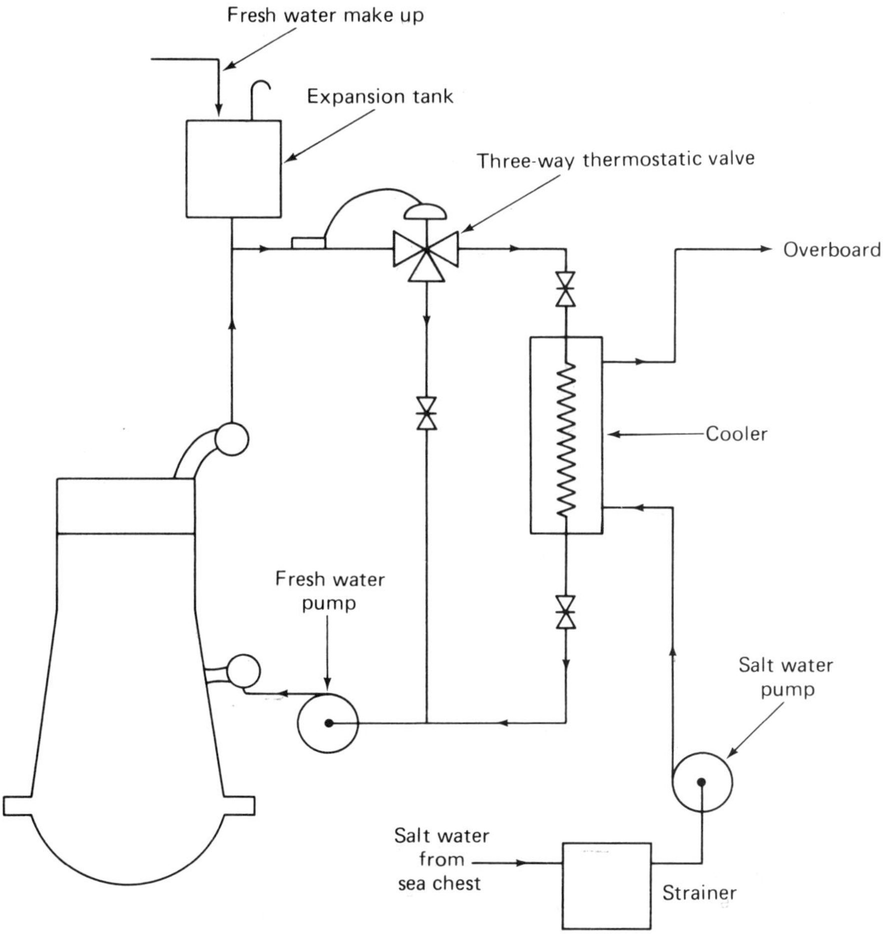

Figure 8-1. Closed cooling water system.

where the engine and the water are warmest. It is considered desirable to have a temperature differential of only 10°F to 20°F across the engine to prevent the buildup of thermal stress. Note that there is an expansion tank above the engine. This tank must have a minimum capacity of 5% to 10% of the engine cooling water by volume. Its purpose is to compensate for the density (and thus, volume) variation in the water due to increases or decreases in temperature. In addition, make-up water for the cooling system is supplied to the expansion tank. A three-way thermostatic valve controls the engine water temperature. The thermostatic valve bypasses the cooler to increase the temperature of the water entering the engine. Bypassing typically occurs during start up and will also occur on a bleed basis during actual engine running as the fresh water engine outlet temperature varies. It is desirable to maintain an average temperature of approximately 160°F to 180°F in the engine. This necessitates mixing the bypassed warm water and the remainder, which is cooled in the heat exchanger, before the mixture reenters the engine. The water cooler uses sea water for cooling, with high and low water suctions through a strainer and salt water pump to the fresh water cooler. There is a manual bypass of the cooler in case the three-way thermostatic valve is broken or if one had never been installed. That one had never been installed is unlikely on current engines.

8.2 Circulating Water Pumps

The circulating water pumps, which force the fresh water through the engine, are of centrifugal design. On direct reversing engines, the pumps are of a modified centrifugal design (with straight vaned impellers), rather than having the curved vane impeller is typical of centrifugal pumps. The modified design reduces pump efficiency but allows the pumps to operate in both ahead and astern directions. On medium and high speed engines, the pumps are driven directly by the engine. Thus, circulation of the water stops when the engine stops. On slow speed diesels, because the complexity of gearing up from the slow speed of the engine to the high speed of the pump, it is more efficient to have the pumps electrically driven. During engine shutdown periods, the pumps should be kept running until the water temperature difference throughout the engine is zero.

8.3 Piston Cooling

Standard practice for many years was to use lube oil for piston cooling, and it is still in practice on many large engines. With the design of the new very large bore engines and the correspondingly large amount of heat that must be removed, however, it was found that water removes heat more efficiently. Since the specific heat of water is about twice that of lube oil, half as much water must be circulated to remove the same amount of heat. Pump size and power can be correspondingly reduced. When water is used for piston cooling, there is a separate water circulating system for the pistons, independent of the main engine cooling water system. Cooling systems may also be found on the fuel injectors to prevent tip overheating. If a tip should overheat, oil contacting it would carbonize, causing coking of the injector nozzle.

8.4 Coolers

The coolers are usually of shell and tube type construction. Some shallow water vessels have a device called a keel cooler. This is a shell and tubing cooler mounted on the outside of the ship's surface. The shell is open to river or fresh water currents so that the outlet from the engine's cooling water passes through tubing in the keel cooler to a pump and back to the engine. The keel cooler relies on either river current or vessel motion to create the water motion on the outside of the tube surface. An advantage of the keel cooler is that it prevents dirt and matter from entering the shell and tube exchanger, a problem that does occur in a conventional set-up.

8.5 Water Conditioning

It is very important that the water used in the engine be distilled or at least softened. Otherwise scale may form, reducing the heat transfer rate, causing overheating of the piston and cylinder, accelerating wear, and, in some instances, causing binding of the rings. It is important that the water cooling systems on engines be kept free of rust, scale, and sludge so that the engine heat may be carried away by the cooling water. Since these contaminants are deposited in a nonuniform manner, they can cause hot spots on cylinder liners or engine blocks. Hot spots result in uneven stresses and can eventually cause material failure. Since an oil film cannot be maintained in an overheated area, rapid piston ring and liner wear may occur. To guard against such conditions, only fresh, soft water that has had inhibitors added should be used as a coolant.

The function of the inhibitors is to deposit a film on the metal surfaces of the cooling system to protect them from corrosion and erosion attack. Erosion is caused by cavitation at the liner or engine surface. This occurs when linear vibration causes vapor bubbles to form; cavitation, in effect, wears away the metal lining. When inhibitor has coated a metal surface, the cavitation effects act on the inhibitor rather than the metal surface. The inhibitor is maintained in the water, so as it is depleted by cavitation it is replaced by chemical coating.

Water Treatment Criteria

To be effective, the cooling system treatment program must meet four conditions:

1. The cooling system should be clean so that the inhibitor will be able to act on the bare metal surfaces of the water jacket.
2. The coolant must be checked periodically for alkalinity, chrome, and chloride content. The inhibitor must be maintained at the strength specified by the manufacturer.
3. The coolant must be clean, mineral-free, soft, fresh water containing a minimum of disolved gases.
4. The cooling system must be mechanically tight to prevent leakage of the water out of the system or of gases into the system.

It is important that hard water and water that is known to be contaminated with chemical or industrial waste be avoided, because these contaminants tend to form sludge and scale within the engine. The chrome inhibitors used for chemical treatment of the water contain no chemical ingredients to soften the water. For engines operating continuously in hard water areas, the installation of water softeners is a viable option. Jacket water softeners are often standard equipment on new diesel engines.

Safety

An important safety consideration is that most corrosion inhibitors are poisonous. Sodium dichromate inhibitors are classified as a health hazard and must be handled carefully. In addition, chromate-type inhibitors are toxic and may produce severe determatitis (skin inflammation). Engine cooling water systems must never be directly connected to the potable water supply. Personnel should avoid any contact of the skin or eyes with chromate-type inhibitors. If chromates come in contact with the skin, the affected areas should be washed with plenty of soap and water immediately after exposure.

The typical cooling water inhibitor is a chromate-alkaline inhibitor consisting of sodium dichromate, sodium hydroxide, and distilled water. When these ingredients are in a solution, the sodium dichromate changes to sodium chromate, which is the active chroming agent. This is the agent that coats the water jacket surfaces.

In ships where the engine cooling water is used as a fresh water evaporator heat source, nitrite borates are often used as part of a total inhibitor system. The inhibitor system is composed of a blend of the nitrite borates and organic compounds. These compounds are not regarded as being as effective as chromates and require high concentrations

(3500–4000 ppm) and a higher alkalinity than chromates. However, nitrite borate has low toxicity, is nonstaining, is compatible with glycol antifreezes, and the crystals, which may form at pump glands, are not as abrasive as chromate crystals.

Tests

There are test kits available to measure the chromate level in the jacket cooling water. Typically this is at least 1000 parts per million (ppm) of chrome, but the engine manufacturer will detail the exact minimum levels of concentration. A concentration level below 550 ppm should not be allowed. It is also important that alkalinity control be maintained. The chrome inhibitor is designed to establish an alkalinity below a pH of 9.8. Should the alkalinity of the jacket cooling water rise above 9.8, the cooling water must be drained and the system refilled. There is no general test for an acidic solution, since it is known that if the concentration of chrome is proper, the solution will have an alkalinity above the minimum required. A third and final test is for chloride content. The inhibitor does not affect the chloride content of the cooling system. Salt water leaking from the heat exchangers is the most common source of chloride. Once any chloride is detected, the system must be drained and repaired. The cooling water system of each engine should be tested weekly to determine the chrome inhibitor level and the chloride and alkalinity levels. When new parts that have not been treated with engine jacket water are installed on an engine, it is recommended that daily chromate tests be made until the chrome concentration in the cooling water stabilizes. This is because the coating of the fresh metal parts will reduce the chromate level and additional treatment with chromate will be necessary.

Alarm System

The alarm system aboard most diesel installations typically senses the water outlet temperature from the engine and will sound if the temperature should increase above a predetermined point.

8.6 Lubricating Oil System

The lubricating systems aboard diesel engines have four main purposes. In order of importance, they are:

1. To reduce friction and wear in the engine.
2. To act as a sealant between the rings and liners.
3. To cool the internal parts of the engine.
4. To clean the engine interior of dirt and carbon.

Base Stock

The lubricating oil used in diesel engines has a high quality base stock to supply the lubricating oil film. Appropriate chemicals are added for particular applications, such as use as cylinder oil and engine oil. The base stock carries and distributes the chemical additives within the oil and provides a heat transfer medium, cooling the pistons and bearings, and sealing the clearance spaces between the piston rings and the cylinder walls. The base stock is normally not one type of oil but rather a blend of oils that meet the viscosity standards required by its use. Generally, a high viscosity neutral distillate is preferable. The base stock has three major qualities: first, high thermal stability, which means the oil film will not carbonize in the hot areas of the cylinder; second, high oxidation stability, which means the oil will not react readily with oxygen; and third, low volatility, which means the oil will not vaporize at the high temperatures encountered in the engine.

Additives

Additives are blended with the base stock and provide the following protection:

1. Alkalinity for acid neutralization. The products of combustion form CO_2 and SO_2, which, in the presence of water, form an acid. The alkaline quality of the oil neutralizes this acid formation.
2. Detergency/dispersency for engine cleanliness. The detergent quality of the oil allows it to absorb the carbon molecules (soot) within the oil film. This prevents a build up within the engine itself.
3. Anti-wear. This property tends to reduce friction between the moving parts.
4. Oxidation resistance. This extends the life of the oil so that it may be used for long periods of time without breaking down.
5. Rust/corrosion protection. The presence of water in the engine will cause the metal surfaces to oxidize. The lubricating oil film should have additives to prevent the oxidation of metal surfaces.
6. Water separation. The oil should have the ability to separate from water, so that they may be easily separated when centrifuged.
7. Anti-foam. The base stock, plus all the other additives, will tend to foam as it is churned within the engine itself. An anti-foam additive prevents this.
8. Pour depressant. This reduces the pour point so that the oil may be used under a wide variety of running temperatures, particularly low temperatures.

Trunk Engine Oil Requirements

Lubricating oils are used in the two basic types of engines: trunk and crosshead. The trunk-type engine requires that the oil provide the following protective services

1. Alkalinity
2. Detergency
3. Thermal stability
4. Oxygen stability
5. Load carrying ability
6. Anti-rust
7. Water separation ability
8. Anti-foam characteristics

The additional characteristic of oil is its load carrying ability. Because of the side thrust of the piston against the cylinder wall, the oil film must remain to deter metal-to-metal contact.

Crosshead Engine Oil Requirements

In the crosshead engine there are two oil systems, one for the engine bearings and moving parts, the other for the cylinder, where the oil comes in contact with combustion gases. In the first case the functions of the engine oil are reduced to providing bearing lubrication, piston cooling, corrosion protection, and promoting cleanliness. Detergent/dispersent additives are not required because there is no contact with the carbon in the exhaust gas. The load carrying ability required is less than that of the oil used in trunk engines because the side thrust developed by the trunk pistons is not present in the crosshead engine. Additive retention requirements are not necessary because of the elimination of the detergency additive. Crosshead engine cylinder oil requirements are more demanding however. The cylinder oils must provide lubrication between the piston and the cylinder to minimize liner and ring wear. They must neutralize the sulfuric and carbonic acids formed by the exhaust products, SO_2 and CO_2, combining with water. In addition, they should maintain engine cleanliness and should not carbonize or coke in the lubricating lines that lead to the cylinder and that are adjacent to the high temperature regions of the cylinder. The properties of crosshead cylinder oils, then, are

1. High alkalinity
2. Thermal stability
3. Oxidation stability

4. Detergency
5. Corrosion protection
6. Anti-wear

Oil Contaminants

There are five common types of contaminants that have a negative affect on the lube oil: combustion products, water, fuel, miscellaneous solids, and other lubricants. Proper operational procedures and maintenance will minimize each of these. One of the common places for water to contaminate lube oil is through the telescoping tubes that provide cooling water to the cylinder. The fuel dilution of lube oil normally occurs through injector malfunction, which allows the fuel to enter the cylinder and wash down into the lube oil system. The products of combustion may blow-by the rings as well as mix with the oil film on the cylinder walls. Solids may enter through blow-by, dirty tanks or lines, or engine repairs. Mixing of different oils is a result of operational error. If crosshead engine oil were mixed with or used in place of the cylinder oil, the resulting mixture would lack some required properties and would be unable to protect the engine.

Shipboard Lube Oil Treatment

To maintain oil purity, there are four general techniques of lube oil treatment used aboard ship. The first is chemical treatment, which is not recommended because it removes additives from the oil. The second is filtration, which removes solids but does not separate water from oil. Water and solids are the major contaminants of lubricating oil. The third, the most commonly used and recommended treatment, is centrifuging. In this case both solids and water are separated from the lube oil by centrifugal force. The fourth, which has a different function from the others, is air blowing. When large amounts of water have mixed with lube oil, the mixture is put in a storage tank. Normally oil and water would be expected to separate but this will not happen if they are in a fine emulsion. Such an emulsion can sit for many days and even be centrifuged without separating. To correct this, warm air, at less than 200°F, is bubbled through the oil from air distribution pipes (drilled pipes) in the bottom of the storage tanks. The water in close proximity to the air will evaporate and be carried away in the dry warm air.

Centrifuging

In centrifuging oils, the throughput of the centrifuge should be about 50% of its maximum capacity if the oil is nondetergent, resulting in a minimum contaminant level as indicated in Figure 8-2. Should the lube oil be of the

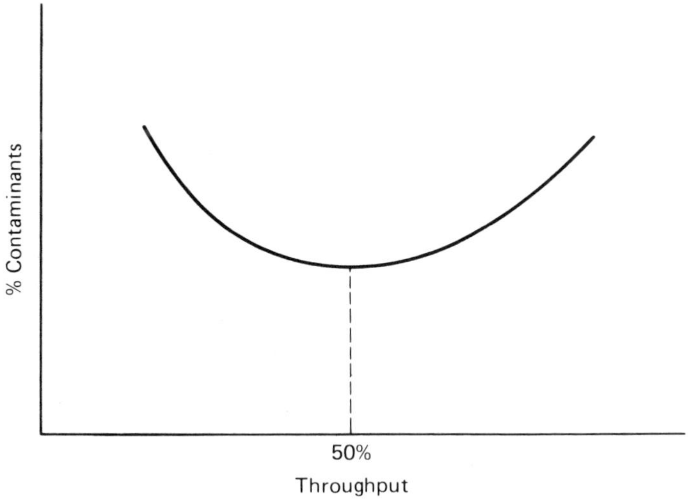

Figure 8-2. Effect of contaminant level as a function of centrifuge capacity.

detergent type, the throughput should be about 25% of the total centrifuging capacity to achieve the same minimum level. The difference is due to the smaller size the contaminant particle found in detergent oils. This is illustrated in Figure 8-3. Because of the small size of the solid particles, separation by centrifugal force does not readily occur and the flowrate through the centrifuge must be lower (slower).

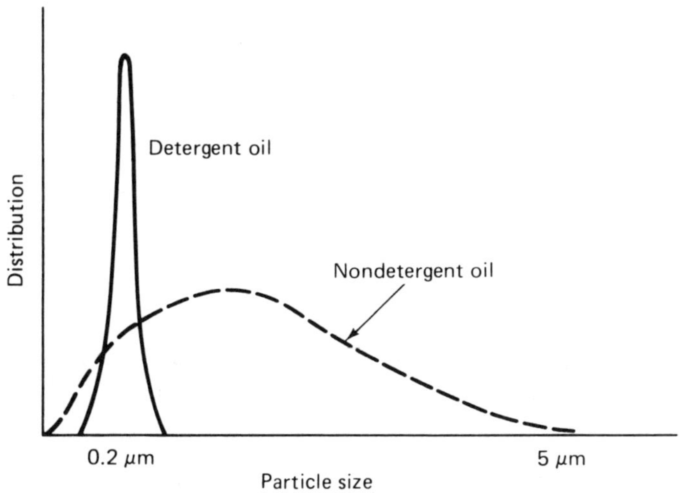

Figure 8-3. Particle size distribution in detergent and nondetergent oils.

Shoreside Analysis

Many of the major oil companies provide an engine oil analysis service, detecting malfunctions in fuel injection equipment, air and oil filters, air blowers, coolant leaks, temperature controls, wear, and corrosion. They analyze changes in oil characteristics and diagnose problems from their analysis. The changes in oil characteristics most commonly come from the depletion of additives due to combustion products; a less common but still frequent occurrence is oil oxidation due to overheating and contaminants. Oxidation typically causes lacquer deposits on the engine surfaces and an increase in oil viscosity due to large molecules being present in the oil. The shorter, more volatile molecules evaporate or will have been destructively distilled.

Oil Sampling

The frequency of oil sampling for analysis varies with the type of engine and with the use of the engine: auxiliary versus main, high and medium speed versus slow speed. A lube oil sampling connection is shown in Figure 8-4. If the pipe is horizontal, the sampling valve should be midway up on the pipe, so that any sludge or deposits that form in the lower half of the pipe would not be included in the oil sample, giving a false value to the oil characteristics. A measure of crankcase oil additive level is noted by its TBN number. The greater the number, the greater the additive level. Normally, if the TBN number is greater than 5, the lube oil is safe to use in an engine. Crankcase oil initially has a TBN number of 70. This is because there is a great amount of sulfur in the fuel, which causes depletion of the additives.

8.7 Purification Systems

There are four types of lubricating oil purification systems found aboard ships. Figure 8-5 illustrates the first, a full flow filter system. The pump takes suction directly from the engine sump, and all the hot oil from the engine passes through the filter, is cooled, and is returned to the engine.

Figure 8-6 illustrates a bypass system. In this case the pump takes suction from the engine, but only 10% of the oil flows through the filter and back to the engine crankcase or sump. The remaining 90% is simply cooled and returned to the engine. The orifice acts as a restrictor to achieve the 10% oil flow. Differential pressure gages are located across the filters to ascertain the filter cleanliness. Should the pressure drop across the filter become greater than the specification, the filter must be cleaned.

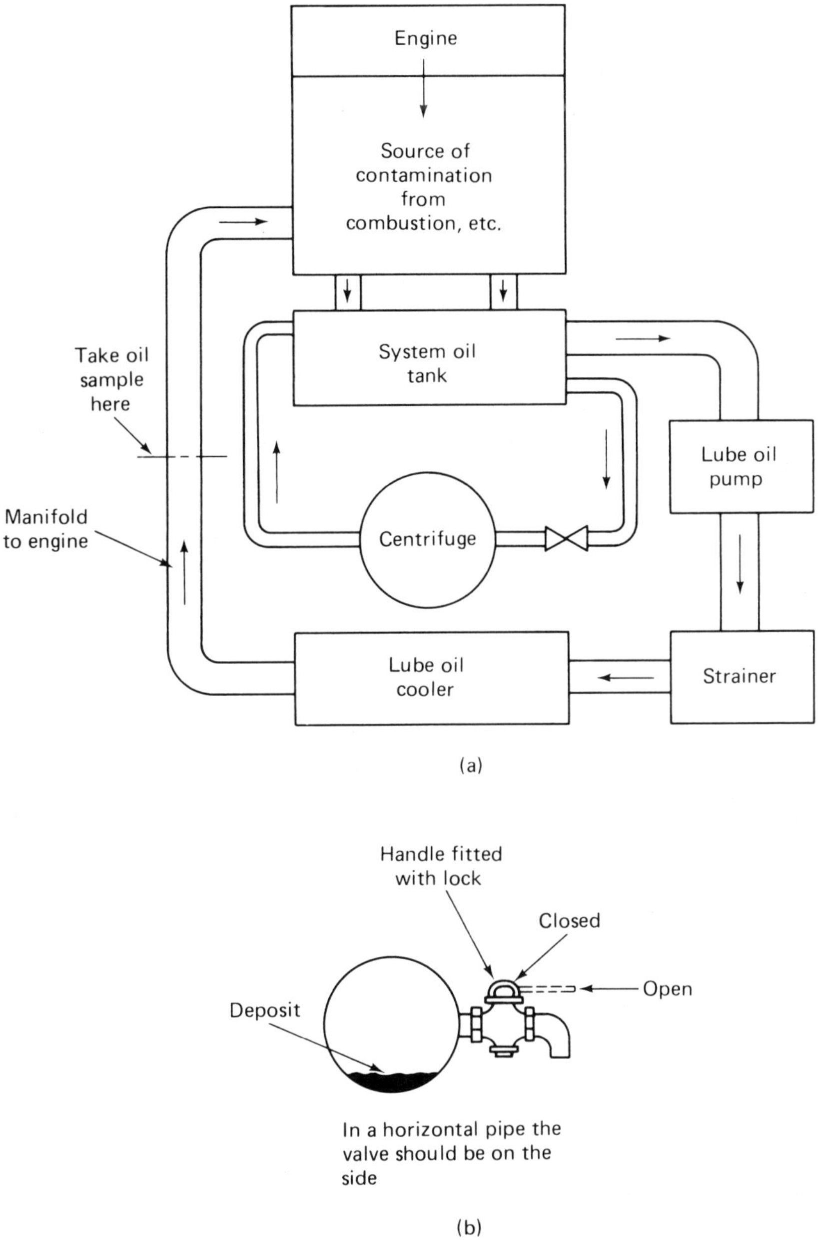

Figure 8-4. Lube oil sampling connection.

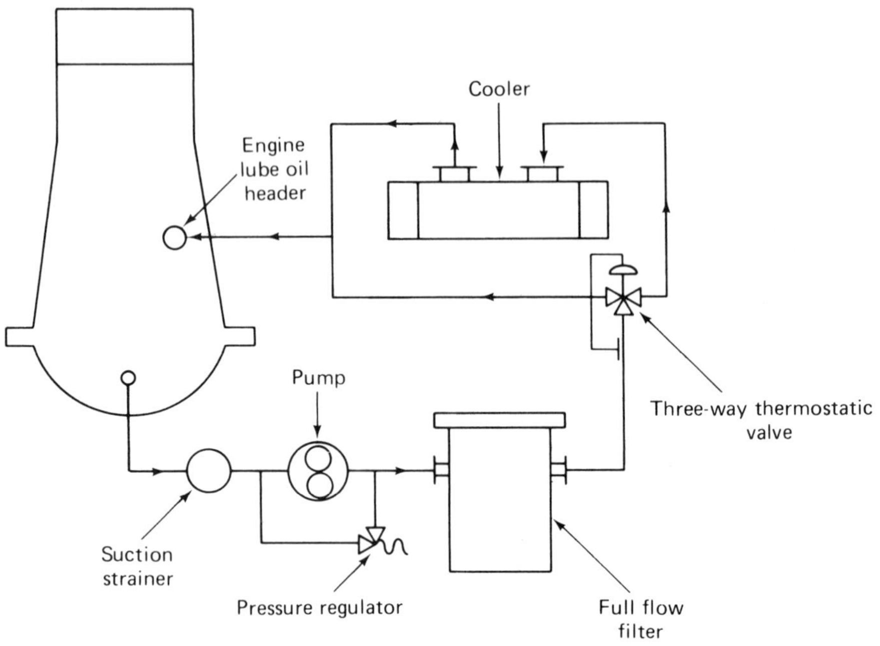

Figure 8-5. Full flow filter system.

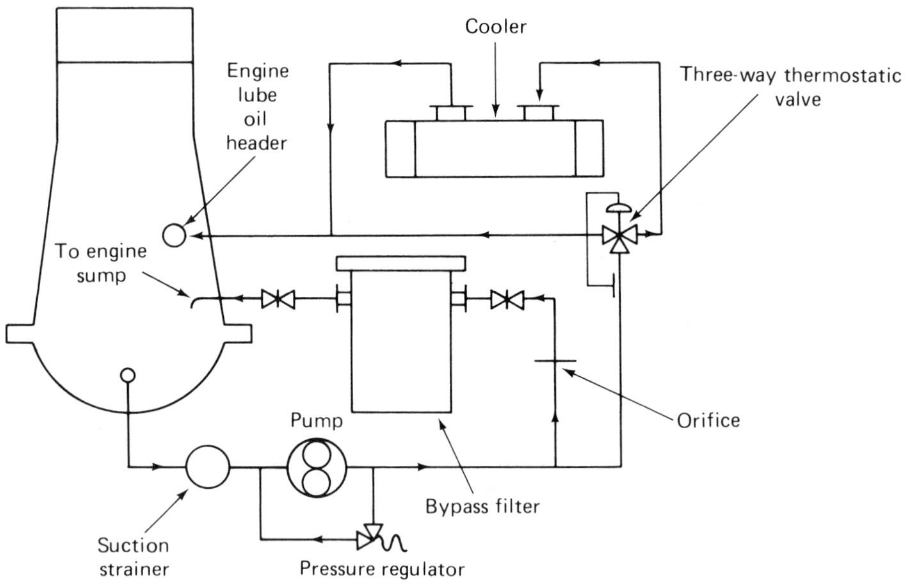

Figure 8-6. Bypass filter system.

The shunt system, not illustrated, is similar to the full flow system with a pressure valve on the filter bypass. This valve allows a constant bypass of hot oil at all times; thus, not all the oil is filtered, but it is all cooled.

The fourth system is a sump system, often used on multiple diesel units driving one common shaft, as illustrated in Figure 8-7. In this system a centrifuge, rather than a filter, is used. The advantage of a centrifuge is that it can remove water, which a filter cannot do. The disadvantage is that it cannot remove fuel. Each engine is centrifuged in turn. This is not a parallel method of operation; first one engine's lube oil is cleaned and then the others.

Figure 8-8 illustrates the batch system, used for cleaning the oil when the engine is stopped, usually while the ship is in port. The dirty oil is pumped from the engine into the dirty oil tank and clean oil is added to the engine. When the vessel has returned to sea, the centrifuge or oil reclaimer is employed to clean the oil. This system may be used in conjunction with any of the previously mentioned lubricating systems, as its purpose is cleaning the oil in batches, not day-to-day cleaning.

The filters are typically of three types: cotton waste, paper, and fullers earth, a diatomaceous earth with the ability to neutralize acidity in the oil.

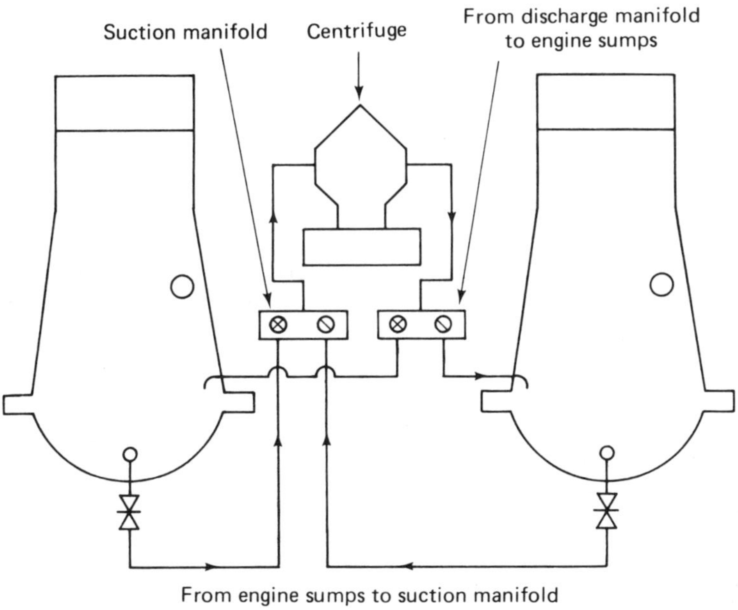

Figure 8-7. Multiengine system.

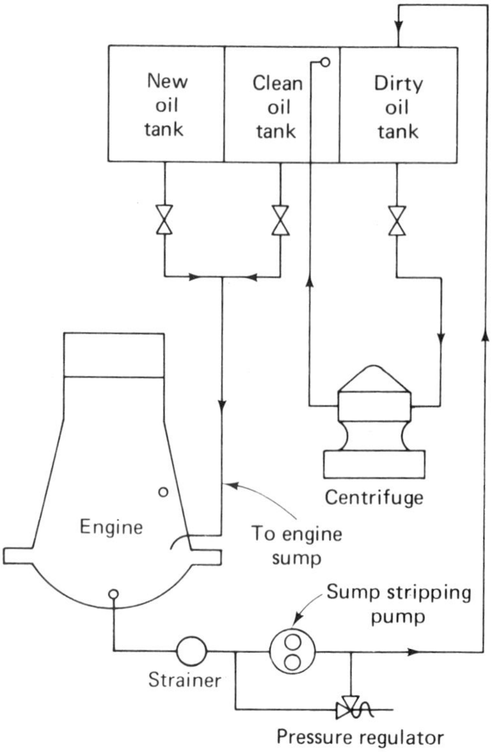

Figure 8-8. Batch system.

8.8 Oil Passages

Figure 8-9 illustrates a crank web with connecting rod, showing the path of the oil as the engine is lubricated. Usually there is a supply line from the main lube oil pressure header to each main bearing. The crankshaft is drilled so the oil will flow from the main bearing to the connecting rod bearing. The oil then passes through another drilled hole to the wrist pin bearing. Excess oil from the wrist pin may spray under the piston crown for additional cooling of the piston, or if there is a separate cooling system for the piston, it will drain back to the crankcase. The lube oil level in the crankcase should be below that of the crankshaft, so the crankshaft itself does not dip into the oil reservoir. Oil leaking from the main and connecting rod bearings is thrown up onto the cylinder walls by rotary motion of the crankshaft, thus lubricates the cylinder walls. This is not splash lubrication in the true sense; that occurs when the crankshaft itself dips into the oil reservoir and oil is thrown by the crank webs.

Where cooling water or lube oil is used for under piston cooling, telescoping tubes usually pass through the fixed wrist pin to the under-crown cooling system. There is one tube on either side of the piston,

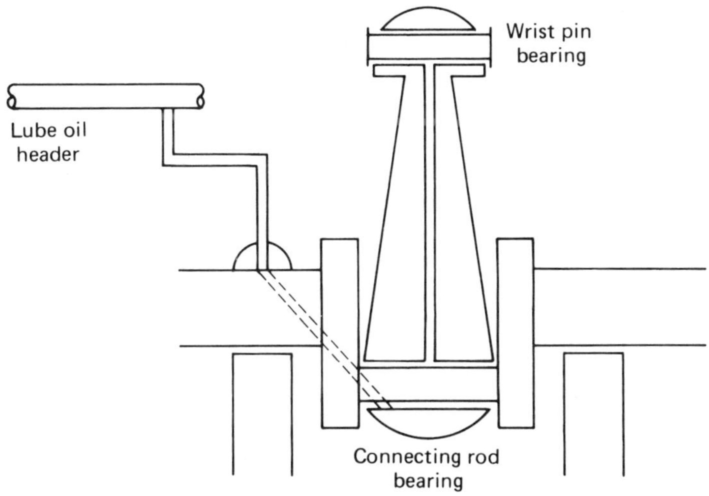

Figure 8-9. Oil passages.

allowing the water or lubricating oil to have a full flow through the underside of the piston. The tubes may be grasshopper tubes, which are balljoint tubes, rather than a tube within a tube. Grasshopper tubes have joints to allow freedom of motion of the tubes and do not depend on maintaining the close clearance required between the telescoping tubes. There is the problem of maintaining the balljoints, however, as they tend to leak.

There are also branch lines from the main lube oil header to the camshaft bearings, to the gear trains, to the rocker arm assembly, and to other moving parts, as the header supplies more than just the main bearing lubricating oil.

8.9 Slow Speed Crosshead Engines

As was noted earlier, two lubricating systems are required in large slow speed crosshead engines. This is because the lubricating oil passes through mechanical force feed lubricators, which are mounted on the engine and provide a small amount of lube oil for each stroke of the engine. There is a small container filled with lube oil mounted below the plunger-type metering pumps. These pumps meter a certain number of drops per minute for each cylinder. Usually there are four to six pumps per cylinder, a figure that varies as a function of the engine bore. The greater the bore, the greater the number of pumps required to provide adequate cylinder linear lubrication. The oil is fed through special fittings that pass through the engine block and screw into holes in the cylinder liner wall. This oil is used only once; thus, there are no oil control rings on crosshead engine pistons. The unburned oil drains back and is bled away. It does not mix with the oil used for crankshaft and bearing lubrication.

8.10 Alarms

There are several alarms located on the pressure lube oil system, to warn of high oil temperature and low oil pressure. The alarms are audible and visible and may have the ability to shut down the engine.

9

Starting and Reversing Systems

Various types of starting and reversing systems are found on diesel engined ships. Large engines are usually started by using air pressure. Other starting systems are used on smaller engines, however, and these will be discussed first. The balance of the chapter on starting will deal with air starting systems.

9.1 Small Engine Starting Systems

Some smaller engines are started electrically. Usually the starting motor is a series-wound dc motor with the voltage across the motor varying between 6 and 120 V. Series-wound motors have a very high starting torque, which, of course, is necessary to move the engine from a stationary position, overcoming high static friction, to a moving state, where the frictional forces are less. Air starting motors, too, may be found on small engines. The air starting motor, attached to the engine through gearing, rotates the engine to start the compression ignition process. There are two types of air starting motors, the piston type and the vane type. This system is quite common on medium speed engines. The third type of small engine starting system is hydraulic. This system is more common than the other two, particularly on small and medium size high speed engines. These starting systems use a piston-type hydraulic motor. The hydraulic pressure in the system is about 3000 psi and the system cut-off pressure, when no more starting is possible, is around 1600 psi. Typically, there is a hand pump and an electrical or engine-driven pump. One of the pumps should be hand driven to allow an emergency startup when there is no power. The system has accumulators, which are charged with nitrogen. As the hydraulic motor is activated, the accumulators maintain a high pressure throughout the system so that there is not an immediate pressure drop.

9.2 Air Starting System

The air starting system, which we will consider in detail, is illustrated in Figure 9-1. A number of requirements are associated with an air starting system. Typically, the air supply in the storage tanks is at a pressure of 500 to 600 psi. This allows a greater mass of air to be stored than if the air were stored at 250 psi, which is normally the required starting air pressure. Note that in the air starting system, one system will supply the auxiliary generator, if necessary, and the main engine starting supply. Also, the ship's service air system may be a reduced air supply off the main engine air starting system. A typical requirement for direct reversing marine propulsion engines is that the air receivers (tanks) hold at least twelve starts of air without the use of a compressor. In case of compressor failure, this allows the ship's master the opportunity to dock or anchor before the engine fails. On ships with controllable pitch propellers, the propulsion engines may run in one direction only. The air receivers in this case need have only six starts of air without the compressor, since these engines need not be stopped so often as do direct reversing engines. A good rule of thumb is that the air capacity in the tanks should be thirty times the mass held in the volume of one engine cylinder at atmospheric pressure. The air receivers, or air tanks, are non-fired pressure vessels and must meet the U. S. Coast Guard requirements in this respect. A relief valve is necessary, and at the lowest part of the tank there must be a drain valve to remove accumulated moisture.

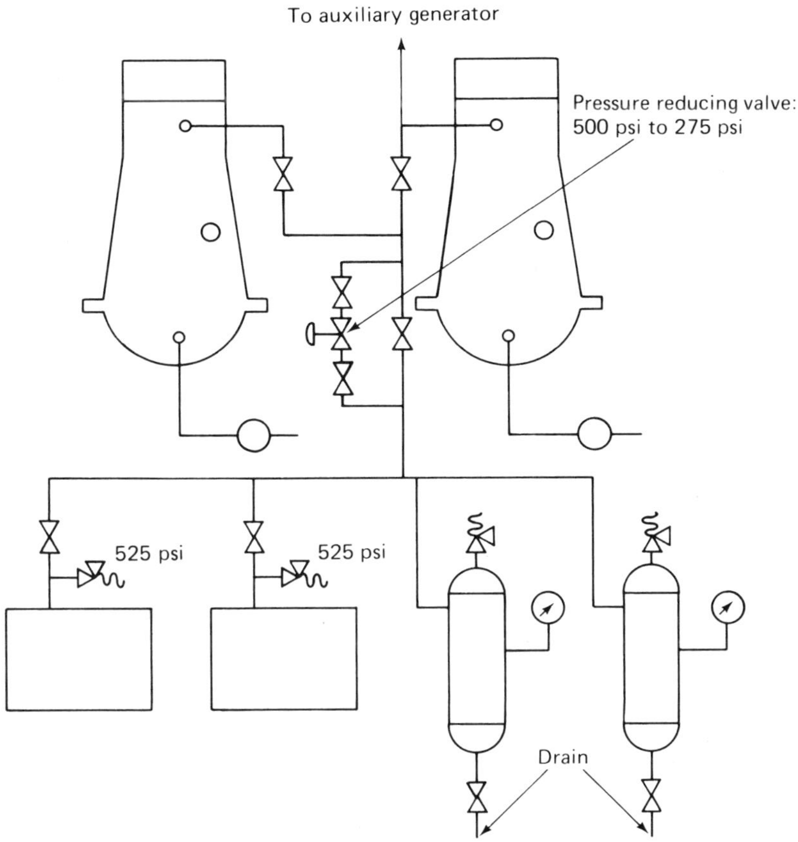

Figure 9-1. An air starting system.

The various types of air starting systems are

1. Mechanical
2. Mechanical with air check valve
3. Pilot operated
   a. Individual
   b. Rotary distributor
4. Distributor (engines with one direction of rotation)

Mechanical

Let us first consider the mechanical air starting system, illustrated in Figure 9-2. As shown in the small diagram at the right of Figure 9-2, the air starting rocker arm is mounted on an eccentric. There is enough

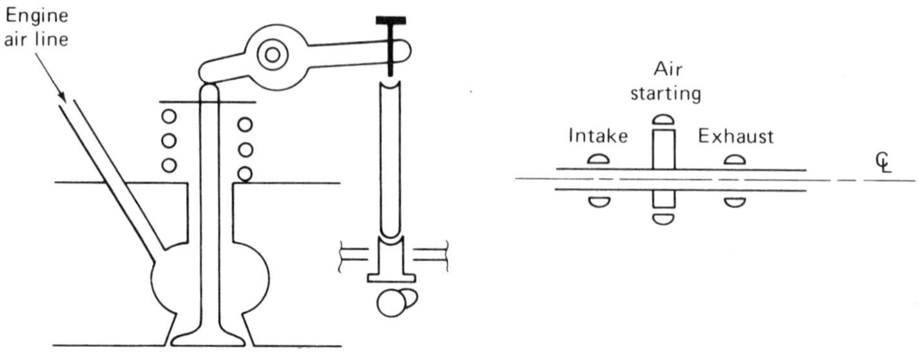

Figure 9-2. Mechanical air starting valve.

clearance so that the cam will not be able to move the rocker arm, hence the valve remains closed. As the camshaft rotates, it cannot engage the air starting rocker arm in this position. When it is necessary to start the engine, the rocker arm support shaft eccentric is lowered. This allows the rocker arm to move towards the air starting valve. Now the cam follower comes in contact with the cam, pushing the rocker arm down, opening the air starting valve, and allowing the starting air to enter the cylinder. As soon as the engine fires, the eccentric is rotated to the upper position, moving the rocker arm away from the valve, and the camshaft can no longer cause enough movement of the rocker arm to engage the valve. This mechanical system is not usually used, but it is the basis from which the other systems were developed. It has a serious disadvantage in that the valve is physically held open by the rocker arm and if fuel should be injected into the system at this time, pressure could build up to a very high level in the air starting lines, causing ruptures or explosions.

Mechanical with Air Check Valve

The mechanical air starting system with an air check valve was developed to overcome these problems. This system is shown in Figure 9-3. The check valve is situated in the engine air line where it terminates in the combustion chamber; this prevents the backflow of air in case a cylinder fires while the main air starting valve is open.

Individual Pilot Operated

The pilot operated system is illustrated in Figure 9-4. The camshaft actuates the pilot valve rather than the large air starting valve in the engine head. A branch line leads from the starting air valve to the pilot valve; another line leads from the starting air valve to the air header. The air valve in the piston is a balanced valve, as shown. Pressure, rather than the

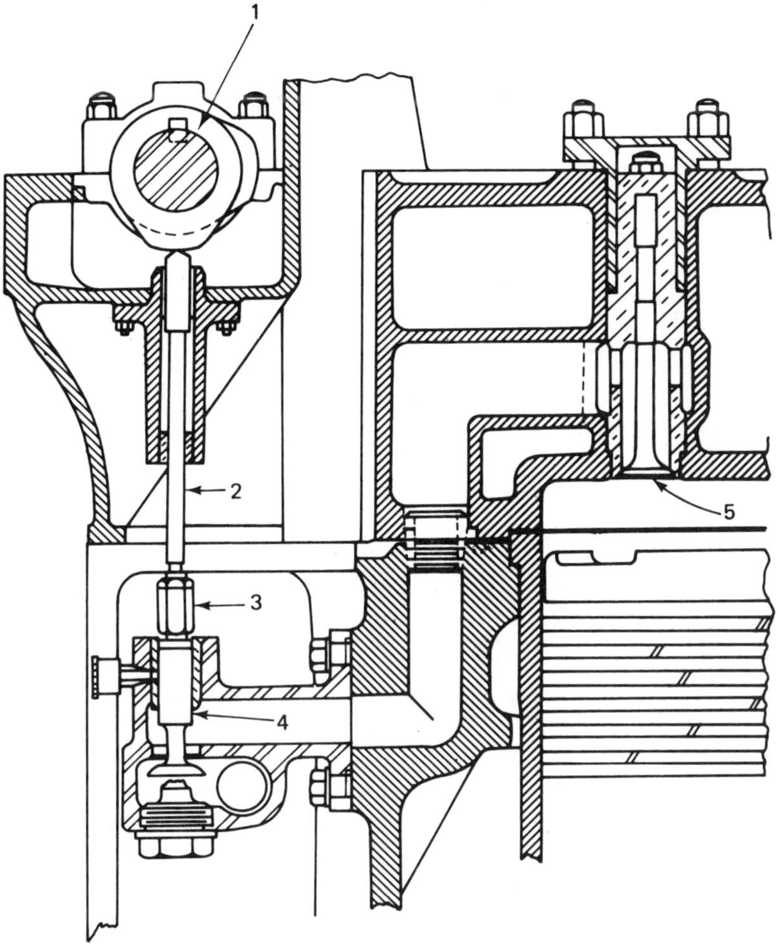

1. Air starting cam
2. Pushrod air start valve
3. Tappet adjustment
4. Air mechanical starting valve
5. Air starting check valve with cage

Figure 9-3. Mechanical air starting valve with air check valve.

rocker arm, opens the valve. In the normal, or closed, position, before the air from the pilot valve enters, the air pressure from the header is essentially balanced. The valve will not open as the valve springs keep it in a closed position. When the camshaft actuates the pilot air valve, air pressure from the header is allowed to go through the pilot valve to the main air starting valve. This causes an unbalanced force because of the increased area on which the pressure is acting downward, and the valve opens. When the cam closes the pilot air valve, the air pressure is no

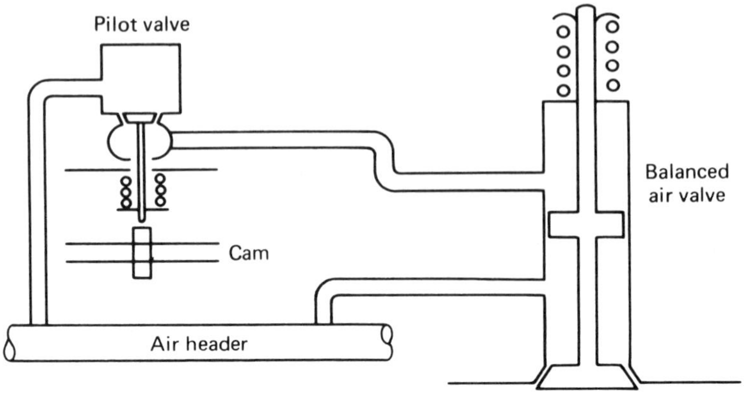

Figure 9-4. Individual pilot operated air valve.

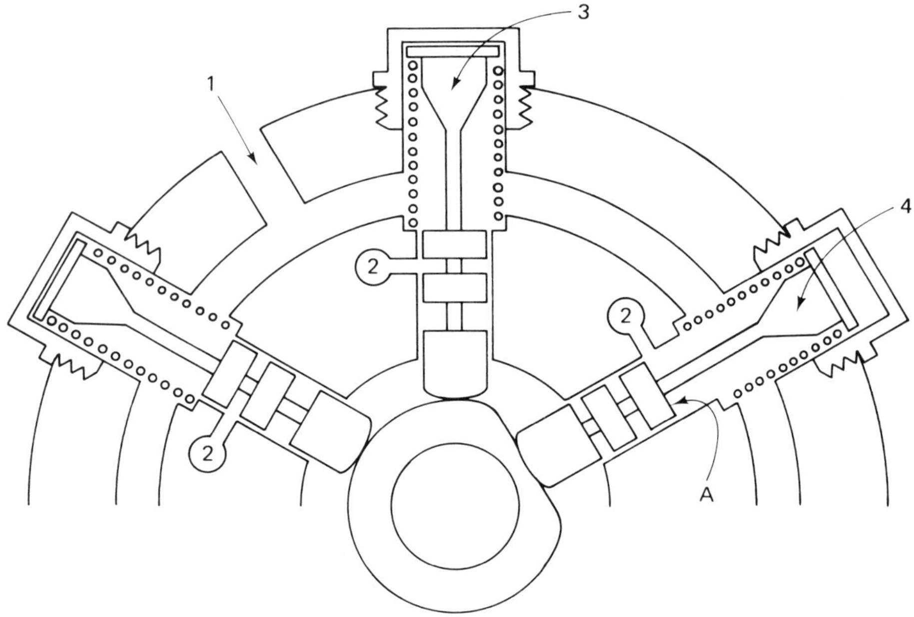

1. Air inlet from starting air header
2. Air outlet to balanced pilot-operated air starting valve
3. Pilot valve in closed position
4. Pilot valve in open position

Note: When air pressure is supplied to the pilot valves, the pressure exerted on land A causes the pilot valve to move down and follow the cam contour. Springs keep the valves off the cam during normal operation.

Figure 9-5. Rotary pilot valve.

longer unbalanced, and the valve closes. Typically one will find $\frac{1}{4}$-in. or $\frac{5}{16}$-in. tubing used for the pilot air lines.

Rotary Distributor Pilot Operated

The rotary distributor pilot system is shown in Figure 9-5. The distributor system uses the same type of balanced piston used and described in the individual pilot operated system. The difference is that, in this case, there is only one cam that operates all of the pilot valves for the engine rather than an individual cam for each starting air valve.

Distributor

Figure 9-6 illustrates a distributor air system. This figure is for a six-cylinder engine which has six starting lines. The air lines on plate 2 are attached to and lead to the individual cylinders. Consider a stopped engine. The air pressure from the air supply line flows through plate 1

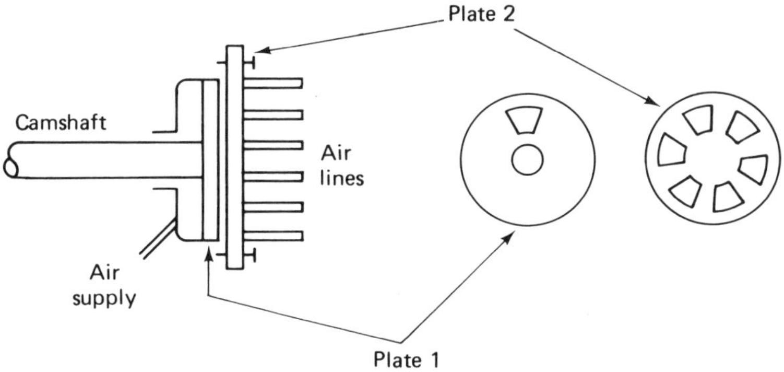

Figure 9-6. Distributor air system.

and plate 2 to the engine. Note that all the air flows from the air supply line. The hole in plate 1 will line up with a hole in plate 2 on the engine starting stroke, allowing air to flow through the plate to the engine. As the engine moves, the camshaft turns, moving plate 1 to line up with another opening on plate 2, thus allowing that cylinder to receive the starting air. The air goes to the starting air valve and into the engine.

9.3 Reversing Systems

For direct propulsion marine engines, there are three basic types of reversing systems. The first is a sliding camshaft; the second, a reversing latch; and the third, a rotating camshaft. The rotating camshaft is used only on

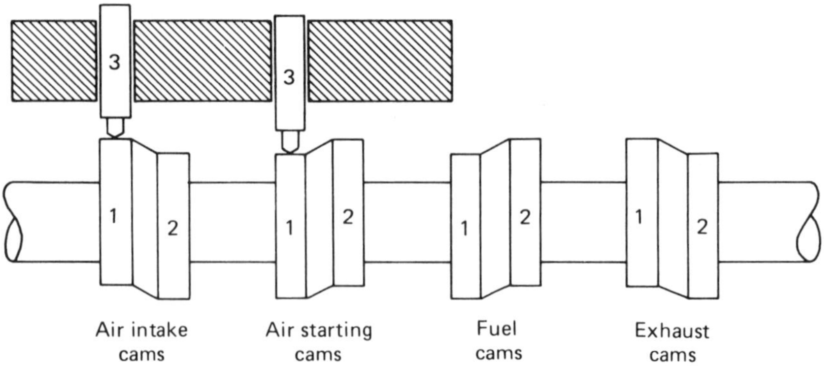

1. Ahead cams
2. Astern cams
3. One cam follower for each set of cams.

Figure 9-7. Sliding camshaft.

two-stroke cycle ported engines. A sliding camshaft is illustrated in Figure 9-7. The rocker arm roller is in constant contact with the cam, either the ahead cam or the astern cam. When the engine stops, the camshaft moves in a lengthwise direction. A ramp between the head and the astern cams allows for constant rocker arm roller contact. There may be cam fol-

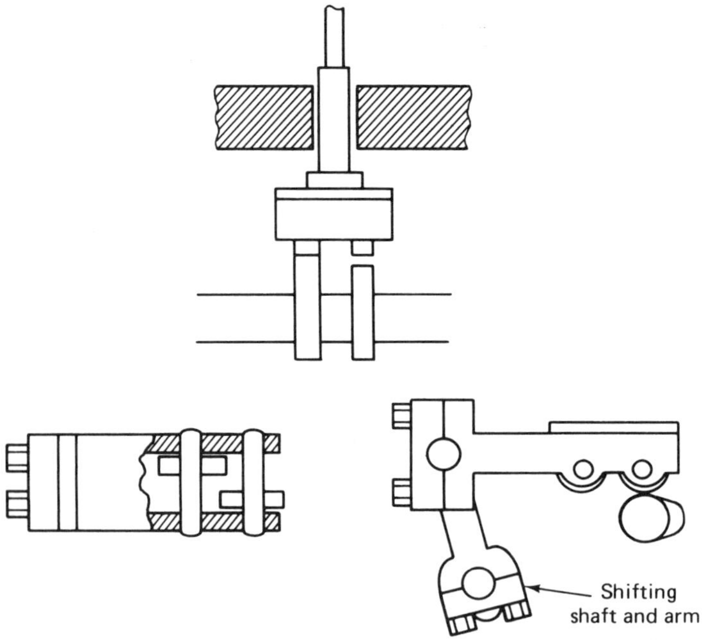

Figure 9-8. Reversing latch.

lowers with pushrods that lead to the rocker arms. This type of reversing system is used on two- and four-stroke cycle engines.

The reversing latch method of moving the camshaft is illustrated in Figure 9-8. The side view in Figure 9-8 shows that there are two rollers on the reversing latch cart. The camshaft does not move, but the reversing latch, as it is called, does. By moving the reversing latch, the roller is allowed to engage with the cam. The cam follower, which would actuate the pushrod, acts in the middle of the reversing block at all times. The reversing latch itself may be either manually or hydraulically actuated. This system is found on two- and four-stroke cycle engines.

In a two-stroke cycle engine, the same fuel cams may be used for ahead and astern operations of the engine. This is accomplished by rotating the camshaft by means of a reversing servomotor, actuated by lubricating oil pressure, as illustrated in Figure 9-9. To reverse the engine, the control lubricating oil is directed to the proper side of the rotary vane, which is connected to the camshaft. The pressure of the oil against the vane then forces the vane to move through an arc to the astern position. The arc of travel is determined by the number of cylinders on the engine.

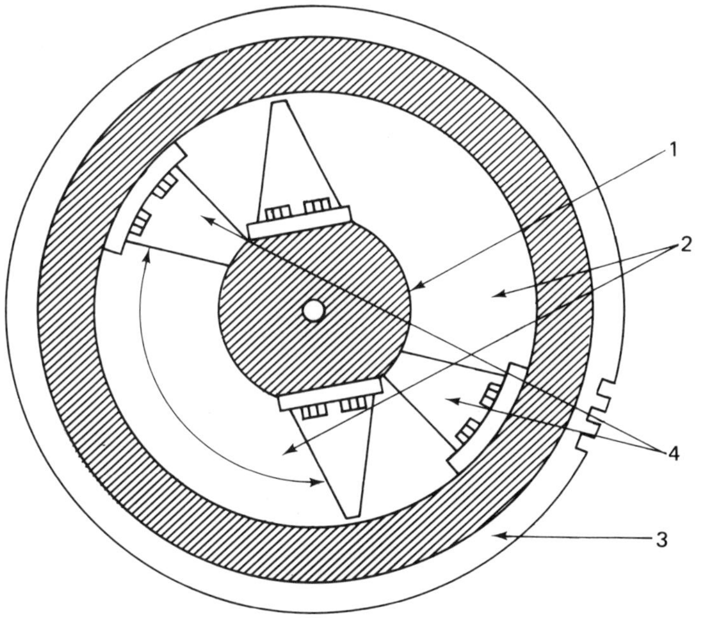

1. Camshaft and attached vanes
2. Oil space for pressure oil to rotate camshaft
3. Cam drive housing and gear
4. Cam stops

Figure 9-9. Rotating camshaft turning mechanism.

To go ahead again the vane is shifted by the lubricating oil pressure back to the ahead position. The same starting air cam is used for both the ahead and the astern positions, and the same fuel cam must also be used. This is possible only if the cams are symmetrical. To prevent oil leakage from the housing, the whole unit is sealed.

9.4 Summary

The reversing system should be able to perform several functions. First, it must stop the fuel to the cylinders when the engine stops. Second, when the engine is put in the ahead position, the fuel pumps should be in a position to supply fuel; then the starting air supply should be activated. Third, when maneuvering from the ahead to the astern position, the fuel must be stopped and the cams moved to the astern position. During this period a brake is often applied to the shaft to stop the engine. If the engine is still rotating, it may continue to run in the same direction even though the cams are shifted. The exhaust valves would then act as inlet valves, and the engine would exhaust through the intake valves into the engine room. As the engine goes to astern, fuel is supplied and then the air is started. There are usually a variety of interlocks associated with engine reversing. If the cam position is not correct, the engine will not start. If the lube oil pressure is not correct, the engine will not start. There may be other interlocks, depending upon the manufacturer.

10

Troubleshooting

Troubleshooting a diesel engine can be viewed as solving problems involving three general categories: trouble with engine operation, e.g., the engine not starting; trouble due to abnormal engine operating conditions while running; and trouble encountered while the engine is being taken apart for routine inspection. Troubleshooting begins by gathering all information observed and information supplied by the operator of the engine. Based on this information the troubleshooter proceeds in a logical fashion through the various symptoms of engine malfunctions. Usually the engine problem does not result in one specific symptom, but by two or more symptoms all of which must be analyzed to lead to the correct problem and its solution. For example, a smoky exhaust may be caused by engine overload, high lube oil consumption, injector malfunction, or some combination of these. By looking at other operating parameters, such as fuel or lube oil consumption, the exact diagnosis as to why the engine is smoking may be deduced. Troubleshooting is a deduction process, solving a mystery based on certain clues.

The engine manual provides a great deal of assistance in troubleshooting a specific engine, as malfunctions indicative of one manufacturer's engines are given attention, as well as those indicative of all engines. Once all information has been gathered, the best approach is to check the troubleshooting charts and proceed by doing the easiest things first. If there are five clues, it may be possible that the one which may be the most easily checked is the one causing the problem, or it may be associated with the most difficult to check. Start with the easiest and hope for the best.

Along with doing the easiest thing first, do not leap to conclusions when starting the troubleshooting task. Gather all information, take your time, think about what has happened, and then proceed in a logical fashion, one step at a time, eliminating various symptoms and finally finding the cause of the problem.

The following are troubleshooting charts which includes symptoms, causes, and remedies for a variety of engine malfunctions. These troubleshooting guides apply to all engines and not to any one specific engine. The problems listed are those that most commonly occur in engine operation.

SYMPTOM	CAUSE	REMEDY
ENGINE FAILS TO START:		
a. ELECTRIC STARTER	discharged battery	charge battery
	defective starting motor or starter solenoid	check continuity of starter system
	poor electrical battery or starter connection	remove and clean connections
b. HYDRAULIC STARTER	low hydraulic oil pressure in accumulator	pump up hydraulic pressure
c. AIR STARTED ENGINE	air line valves not open to engine	check valves in starting air system
	too low air starting pressure	start compressors to build up pressure
	improper air starting valve timing	check air starting timing

SYMPTOM	CAUSE	REMEDY
FAILURE TO START	fuel shut off	open proper valves from fuel tank
	service or fuel tank empty	refill tank
	fuel injection system air bound	prime fuel system
	clogged air filter	replace air filter
	engine cranks over too slowly	check for weak battery; starting air pressure may be too low
	improper fuel metering pump timing	check pump timing
	loose high pressure fuel line connections from injection pump to injector	tighten connection
	improper intake and exhaust valve clearances	readjust valve clearance

SYMPTOM	CAUSE	REMEDY
HARD STARTING	fuel pump needs priming	prime system
	air in system	prime system
	incorrect fuel pump timing	check pump timing
	throttle not set to starting fuel position	move throttle to start position
	defective injectors	test and adjust or replace
	water in fuel system	check system for water
	leaky intake or exhaust valves	check engine compression
	improper valve clearance	check intake and exhaust valve clearances
	engine too cold	

SYMPTOM	CAUSES	REMEDY
ENGINE DOES NOT COME UP TO SPEED OR DEVELOP FULL POWER; ENGINE LOSES POWER	clogged air filter	replace air filter
	clogged fuel filters	check and replace
	insufficient oil pressure at fuel injection pumps	check to see if booster pump is operating,
		check header pressure regulator
	improper fuel injector pump timing	check fuel pump timing
	water in fuel	
	air in fuel injection system	bleed system to injectors
	low compression	check cylinder compression pressure
	governor not functioning properly	check governor adjustments
	injectors are defective	test injectors and adjust or replace
	cylinders misfiring	check for air in system
		check for water in system
		check for valve clearances
		faulty injector
		loose high pressure fuel line connection
	collapsed or clogged exhaust muffler	check exhaust back pressure
	poor quality fuel	

SYMPTOM	CAUSE	REMEDY
ENGINE RUNNING UNEVENLY	improperly balanced fuel injection system	check metering pump fuel rack settings
	defective injector	test injectors and adjust or replace
	malfunctioning hydraulic governor	adjust compensating needle
	air leaks in booster or transfer suction line	check for air leakage in fuel system

SYMPTOM	CAUSE	REMEDY
DETONATION	early injection timing	check and adjust timing
	defective fuel injectors	test injectors
	poor quality fuel	
	engine overload	remove some of the engine load

SYMPTOM	CAUSE	REMEDY
ENGINE OVERHEATS	lack of coolant in system	check radiator or expansion tank
	fan belt loose	check fan belt tension and adjust
	radiator or coolers are dirty	clean and replace
	lack of lubricating oil causing friction	check oil sump
	worn water pump	replace pump
	engine overload	remove excessive load
	late fuel injection timing	check injection fuel pump timing
	faulty thermostat or thermostatic valve	replace thermostat or adjust valve

SYMPTOM	CAUSE	REMEDY
EXCESSIVE LUBRICATING OIL CONSUMPTION	external engine oil leaks	locate and replace gaskets, etc.
	excessive valve guide clearances	replace valve guides
	worn rings	check compression pressure
	excessively worn cylinder liners	check compression pressure
	excessive main and connecting rod bearing clearances	check bearing clearances
	excessively high oil temperature	clogged or dirty oil cooler

SYMPTOM	CAUSE	REMEDY
LOW LUBRICATING OIL PRESSURE	dirty or clogged oil filters	check and replace
	fuel oil dilution	locate internal fuel leaks and change lube oil
	faulty oil pressure regulator	check and reset
	worn lube oil pump	
	too light an oil for high temperature operation	change to heavier grade
	excessively worn main and connecting rod bearings	

SYMPTOM	CAUSE	REMEDY
BLACK SMOKE	engine overload	
	malfunctioning injectors	test, repair, or replace
	improper fuel injection timing	check injector timing
	unbalanced fuel injection pumps	check rack settings for each pump
	dirty oil filter	replace filter
	low engine compression	run cylinder compression pressure test
BLUE SMOKE	excessive lube oil in combus tion space	
	worn valve guides	replace guides or seats
	worn oil control rings	replace rings
	worn cylinder liner	replace liner
	excessive main and connect- ing rod bearing clearances	replace bearing
WHITE SMOKE	engine misfiring	
	air in fuel injection system	prime fuel injection system
	water in fuel	drain and replace fuel
	bad injector	test, repair, or replace

Appendices

APPENDIX A-1 DEGREES API WITH CORRESPONDING SPECIFIC GRAVITY, BARRELS PER TON, AND POUNDS PER GALLON, ALL AT 60°F.

Degrees API Gravity at 60°F	Specific Gravity at 60°F	Barrels at 60°F per Long Ton	Pounds per Gallon at 60°F
0.0	1.076046	5.95151	8.96131
0.5	1.071970	5.97414	8.92736
1.0	1.067925	5.99677	8.89368
1.5	1.063910	6.01940	8.86024
2.0	1.059926	6.04203	8.82706
2.5	1.055971	6.06466	8.79412
3.0	1.052045	6.08729	8.76143
3.5	1.048149	6.10992	8.72898
4.0	1.044281	6.13255	8.69677
4.5	1.040442	6.15518	8.66479
5.0	1.036631	6.17781	8.63305
5.5	1.032847	6.20043	8.60155
6.0	1.029091	6.22306	8.57027
6.5	1.025363	6.24569	8.53922
7.0	1.021661	6.26832	8.50839
7.5	1.017986	6.29095	8.47778
8.0	1.014337	6.31358	8.44740
8.5	1.010715	6.33621	8.41723
9.0	1.007118	6.35884	8.38727
9.5	1.003547	6.38147	8.35753
10.0	1.000000	6.40410	8.32800
10.5	0.996479	6.42673	8.29868
11.0	0.992982	6.44936	8.26956
11.5	0.989510	6.47199	8.24064
12.0	0.986063	6.49462	8.21193
12.5	0.982639	6.51725	8.18342
13.0	0.979239	6.53987	8.15510
13.5	0.975862	6.56250	8.12698
14.0	0.972509	6.58513	8.09905
14.5	0.969178	6.60776	8.07132
15.0	0.965870	6.63039	8.04377
15.5	0.962585	6.65302	8.01641
16.0	0.959322	6.67565	7.98923
16.5	0.956081	6.69828	7.96224
17.0	0.952862	6.72091	7.93543
17.5	0.949664	6.74354	7.90881
18.0	0.946488	6.76617	7.88235
18.5	0.943333	6.78880	7.85608
19.0	0.940199	6.81143	7.82998
19.5	0.937086	6.83406	7.80405
20.0	0.933993	6.85669	7.77830
20.5	0.930921	6.87931	7.75271

APPENDIX A-1 *(Continued)*.

Degrees API Gravity at 60°F	Specific Gravity at 60°F	Barrels at 60°F per Long Ton	Pounds per Gallon at 60°F
21.0	0.927869	6.90194	7.72729
21.5	0.924837	6.92457	7.70204
22.0	0.921824	6.94720	7.67695
22.5	0.918831	6.96983	7.65203
23.0	0.915858	6.99246	7.62726
23.5	0.912903	7.01509	7.60266
24.0	0.909968	7.03772	7.57821
24.5	0.907051	7.06035	7.55392
25.0	0.904153	7.08298	7.52979
25.5	0.901274	7.10561	7.50581
26.0	0.898413	7.12824	7.48198
26.5	0.895570	7.15087	7.45830
27.0	0.892744	7.17350	7.43478
27.5	0.889937	7.19612	7.41140
28.0	0.887147	7.21875	7.38816
28.5	0.884375	7.24138	7.36507
29.0	0.881620	7.26401	7.34213
29.5	0.878882	7.28664	7.31933
30.0	0.876161	7.30927	7.29667
30.5	0.873457	7.33190	7.27415
31.0	0.870769	7.35453	7.25177
31.5	0.868098	7.37716	7.22952
32.0	0.865443	7.39979	7.20741
32.5	0.862805	7.42242	7.18544
33.0	0.860182	7.44505	7.16360
33.5	0.857576	7.46768	7.14189
34.0	0.854985	7.49031	7.12031
34.5	0.852410	7.51294	7.09887
35.0	0.849850	7.53556	7.07755
35.5	0.847305	7.55819	7.05636
36.0	0.844776	7.58082	7.03530
36.5	0.842262	7.60345	7.01436
37.0	0.839763	7.62608	6.99354
37.5	0.837278	7.64871	6.97285
38.0	0.834808	7.67134	6.95228
38.5	0.832353	7.69397	6.93184
39.0	0.829912	7.71660	6.91151
39.5	0.827485	7.73923	6.89130
40.0	0.825073	7.76186	6.87121
40.5	0.822674	7.78449	6.85123
41.0	0.820290	7.80712	6.83137
41.5	0.817919	7.82975	6.81163
42.0	0.815562	7.85238	6.79200

APPENDIX A-1 (*Continued*).

Degrees API Gravity at 60°F	Specific Gravity at 60°F	Barrels at 60°F per Long Ton	Pounds per Gallon at 60°F
42.5	0.813218	7.87500	6.77248
43.0	0.810888	7.89763	6.75308
43.5	0.808571	7.92026	6.73378
44.0	0.806268	7.94289	6.71460
44.5	0.803977	7.96552	6.69552
45.0	0.801700	7.98815	6.67656
45.5	0.799435	8.01078	6.65769
46.0	0.797183	8.03341	6.63894
46.5	0.794944	8.05604	6.62029
47.0	0.792717	8.07867	6.60175
47.5	0.790503	8.10130	6.58331
48.0	0.788301	8.12393	6.56497
48.5	0.786111	8.14656	6.54673
49.0	0.783934	8.16919	6.52860
49.5	0.781768	8.19182	6.51056
50.0	0.779614	8.21444	6.49263

APPENDIX A.2a FUELS WITH API GRAVITY 0-14.9 AT 60° F

T	m	T	m	T	m	T	m	T	m
0	1.0211	56	1.0014	112	0.9819	168	0.9628	224	0.9439
1	1.0208	57	1.0010	113	0.9816	169	0.9624	225	0.9436
2	1.0204	58	1.0007	114	0.9813	170	0.9621	226	0.9432
3	1.0201	59	1.0003	115	0.9809	171	0.9618	227	0.9429
4	1.0197	60	1.0000	116	0.9806	172	0.9614	228	0.9426
5	1.0194	61	0.9997	117	0.9802	173	0.9611	229	0.9422
6	1.0190	62	0.9993	118	0.9799	174	0.9607	230	0.9419
7	1.0186	63	0.9990	119	0.9795	175	0.9604	231	0.9416
8	1.0183	64	0.9986	120	0.9792	176	0.9601	232	0.9412
9	1.0179	65	0.9983	121	0.9788	177	0.9597	233	0.9409
10	1.0176	66	0.9979	122	0.9785	178	0.9594	234	9.9405
11	1.0172	67	0.9976	123	0.9782	179	0.9590	235	0.9402
12	1.0169	68	0.9972	124	0.9778	180	0.9587	236	0.9399
13	1.0165	69	0.9969	125	0.9775	181	0.9584	237	0.9395
14	1.0162	70	0.9965	126	0.9771	182	0.9580	238	0.9392
15	1.0158	71	0.9962	127	0.9768	183	0.9577	239	0.9389
16	1.0155	72	0.9958	128	0.9764	184	0.9574	240	0.9385
17	1.0151	73	0.9955	129	0.9761	185	0.9570	241	0.9382
18	1.0148	74	0.9951	130	0.9758	186	0.9567	242	0.9379
19	1.0144	75	0.9948	131	0.9754	187	0.9563	243	0.9375
20	1.0141	76	0.9944	132	0.9751	188	0.9560	244	0.9372
21	1.0137	77	0.9941	133	0.9747	189	0.9557	245	0.9369
22	1.0133	78	0.9937	134	0.9744	190	0.9553	246	0.9365
23	1.0130	79	0.9934	135	0.9740	191	0.9550	247	0.9362
24	1.0126	80	0.9930	136	0.9737	192	0.9547	248	0.9359
25	1.0123	81	0.9927	137	0.9734	193	0.9543	249	0.9356
26	1.0119	82'	0.9923	138	0.9730	194	0.9540	250	0.9352
27	1.0116	83	0.9920	139	0.9727	195	0.9536	251	0.9349
28	1.0112	84	0.9916	140	0.9723	196	0.9533	252	0.9346
29	1.0109	85	0.9913	141	0.9720	197	0.9530	253	0.9342
30	1.0105	86	0.9909	142	0.9716	198	0.9526	254	0.9339
31	1.0102	87	0.9906	143	0.9713	199	0.9523	255	0.9336
32	1.0098	88	0.9902	144	0.9710	200	0.9520	256	0.9332
33	1.0095	89	0.9899	145	0.9706	201	0.9516	257	0.9329
34	1.0091	90	0.9896	146	0.9703	202	0.9513	258	0.9326
35	1.0088	91	0.9892	147	0.9699	203	0.9509	259	0.9322
36	1.0084	92	0.9889	148	0.9696	204	0.9506	260	0.9319
37	1.0081	93	0.9885	149	0.9693	205	0.9503	261	0.9316
38	1.0077	94	0.9882	150	0.9689	206	0.9499	262	0.9312
39	1.0074	95	0.9878	151	0.9686	207	0.9496	263	0.9309
40	1.0070	96	0.9875	152	0.9682	208	0.9493	264	0.9306

T = temperature in degrees Fahrenheit
m = multiplier
The multiplier, m, will reduce the volume of fuel at any temperature T to its equivalent volume at 60° F.

APPENDIX A-2a (*Continued*).

T	m	T	m	T	m	T	m	T	m
41	1.0067	97	0.9871	153	0.9679	209	0.9489	265	0.9302
42	1.0063	98	0.9868	154	0.9675	210	0.9486	266	0.9299
43	1.0060	99	0.9864	155	0.9672	211	0.9483	267	0.9296
44	1.0056	100	0.9861	156	0.9669	212	0.9479	268	0.9293
45	1.0053	101	0.9857	157	0.9665	213	0.9476	269	0.9289
46	1.0049	102	0.9854	158	0.9662	214	0.9472	270	0.9286
47	1.0046	103	0.9851	159	0.9658	215	0.9469	271	0.9283
48	1.0042	104	0.9847	160	0.9655	216	0.9466	272	0.9279
49	1.0038	105	0.9844	161	0.9652	217	0.9462	273	0.9276
50	1.0035	106	0.9840	162	0.9648	218	0.9459	274	0.9273
51	1.0031	107	0.9837	163	0.9645	219	0.9456	275	0.9269
52	1.0028	108	0.9833	164	0.9641	220	0.9452	276	0.9266
53	1.0024	109	0.9830	165	0.9638	221	0.9449	277	0.9263
54	1.0021	110	0.9826	166	0.9635	222	0.9446	278	0.9259
55	1.0017	111	0.9823	167	0.9631	223	0.9442	—	—

T = temperature in degrees Fahrenheit
m = multiplier
The multiplier, m, will reduce the volume of fuel at any temperature T to its equivalent volume at 60° F.

APPENDIX A-2b FUELS WITH API GRAVITY 15.0–34.9 AT 60° F

T	m	T	m	T	m	T	m	T	m
0	1.0241	56	1.0016	112	0.9795	168	0.9578	224	0.9365
1	1.0237	57	1.0012	113	0.9791	169	0.9574	225	0.9361
2	1.0233	58	1.0008	114	0.9787	170	0.9570	226	0.9358
3	1.0229	59	1.0004	115	0.9783	171	0.9566	227	0.9354
4	1.0225	60	1.0000	116	0.9779	172	0.9562	228	0.9350
5	1.0221	61	0.9996	117	0.9775	173	0.9559	229	0.9346
6	1.0217	62	0.9992	118	0.9771	174	0.9555	230	0.9343
7	1.0213	63	0.9988	119	0.9767	175	0.9551	231	0.9339
8	1.0209	64	0.9984	120	0.9763	176	0.9547	232	0.9335
9	1.0205	65	0.9980	121	0.9760	177	0.9543	233	0.9331
10	1.0201	66	0.9976	122	0.9756	178	0.9539	234	0.9328
11	1.0197	67	0.9972	123	0.9752	179	0.9536	235	0.9324
12	1.0193	68	0.9968	124	0.9748	180	0.9532	236	0.9320
13	1.0189	69	0.9964	125	0.9744	181	0.9528	237	0.9316
14	1.0185	70	0.9960	126	0.9740	182	0.9524	238	0.9313
15	1.0181	71	0.9956	127	0.9736	183	0.9520	239	0.9309
16	1.0177	72	0.9952	128	0.9732	184	0.9517	240	0.9305
17	1.0173	73	0.9948	129	0.9728	185	0.9513	241	0.9301
18	1.0168	74	0.9944	130	0.9725	186	0.9509	242	0.9298
19	1.0164	75	0.9940	131	0.9721	187	0.9505	243	0.9294
20	1.0160	76	0.9936	132	0.9717	188	0.9501	244	0.9290
21	1.0156	77	0.9932	133	0.9713	189	0.9498	245	0.9286
22	1.0152	78	0.9929	134	0.9709	190	0.9494	246	0.9283
23	1.0148	79	0.9925	135	0.9705	191	0.9490	247	0.9279
24	1.0144	80	0.9921	136	0.9701	192	0.9486	248	0.9275
25	1.0140	81	0.9917	137	0.9697	193	0.9482	249	0.9272
26	1.0136	82	0.9913	138	0.9693	194	0.9478	250	0.9268
27	1.0132	83	0.9909	139	0.9690	195	0.9475	251	0.9264
28	1.0128	84	0.9905	140	0.9686	196	0.9471	252	0.9260
29	1.0124	85	0.9901	141	0.9682	197	0.9467	253	0.9257
30	1.0120	86	0.9897	142	0.9678	198	0.9463	254	0.9253
31	1.0116	87	0.9893	143	0.9674	199	0.9460	255	0.9249
32	1.0112	88	0.9889	144	0.9670	200	0.9456	256	0.9245
33	1.0108	89	0.9885	145	0.9666	201	0.9452	257	0.9242
34	1.0104	90	0.9881	146	0.9662	202	0.9448	258	0.9238
35	1.0100	91	0.9877	147	0.9659	203	0.9444	259	0.9234
36	1.0096	92	0.9873	148	0.9655	204	0.9441	260	0.9231
37	1.0092	93	0.9869	149	0.9651	205	0.9437	261	0.9227
38	1.0088	94	0.9865	150	0.9647	206	0.9433	262	0.9223
39	1.0084	95	0.9861	151	0.9643	207	0.9429	263	0.9219
40	1.0080	96	0.9857	152	0.9639	208	0.9425	264	0.9216

T = temperature (° F)
m = multiplier
The multiplier, m, will reduce the volume of fuel at any temperature T to its equivalent volume at 60° F.

APPENDIX A-2b (*Continued*).

T	*m*	*T*	*m*	*T*	*m*	*T*	*m*	*T*	*m*
41	1.0076	97	0.9854	153	0.9635	209	0.9422	265	0.9212
42	1.0072	98	0.9850	154	0.9632	210	0.9418	266	0.9208
43	1.0068	99	0.9846	155	0.9628	211	0.9414	267	0.9205
44	1.0064	100	0.9842	156	0.9624	212	0.9410	268	0.9201
45	1.0060	101	0.9838	157	0.9620	213	0.9407	269	0.9197
46	1.0056	102	0.9834	158	0.9616	214	0.9403	270	0.9194
47	1.0052	103	0.9830	159	0.9612	215	0.9399	271	0.9190
48	1.0048	104	0.9826	160	0.9609	216	0.9395	272	0.9186
49	1.0044	105	0.9822	161	0.9605	217	0.9391	273	0.9182
50	1.0040	106	0.9818	162	0.9601	218	0.9388	274	0.9179
51	1.0036	107	0.9814	163	0.9597	219	0.9384	275	0.9175
52	1.0032	108	0.9810	164	0.9593	220	0.9380	276	0.9171
53	1.0028	109	0.9806	165	0.9589	221	0.9376	277	0.9168
54	1.0024	110	0.9803	166	0.9585	222	0.9373	278	0.9164
55	1.0020	111	0.9799	167	0.9582	223	0.9369	—	—

APPENDIX A-2c FUELS WITH API GRAVITY 35.0–50.9 AT 60°F

T	*m*	*T*	*m*	*T*	*m*	*T*	*m*	*T*	*m*
0	1.0298	30	1.0149	60	1.0000	90	0.9851	120	0.9702
1	1.0293	31	1.0144	61	0.9995	91	0.9846	121	0.9697
2	1.0288	32	1.0139	62	0.9990	92	0.9841	122	0.9692
3	1.0283	33	1.0134	63	0.9985	93	0.9836	123	0.9687
4	1.0278	34	1.0129	64	0.9980	94	0.9831	124	0.9682
5	1.0273	35	1.0124	65	0.9975	95	0.9826	125	0.9677
6	1.0268	36	1.0119	66	0.9970	96	0.9821	126	0.9672
7	1.0263	37	1.0114	67	0.9965	97	0.9816	127	0.9667
8	1.0258	38	1.0109	68	0.9960	98	0.9811	128	0.9662
9	1.0253	39	1.0104	69	0.9955	99	0.9806	129	0.9657
10	1.0248	40	1.0099	70	0.9950	100	0.9801	130	0.9652
11	1.0243	41	1.0094	71	0.9945	101	0.9796	131	0.9647
12	1.0238	42	1.0089	72	0.9940	102	0.9791	132	0.9642
13	1.0233	43	1.0084	73	0.9935	103	0.9786	133	0.9637
14	1.0228	44	1.0079	74	0.9930	104	0.9781	134	0.9632
15	1.0223	45	1.0075	75	0.9925	105	0.9776	135	0.9627
16	1.0218	46	1.0070	76	0.9920	106	0.9771	136	0.9622
17	1.0214	47	1.0065	77	0.9916	107	0.9766	137	0.9617
18	1.0209	48	1.0060	78	0.9911	108	0.9761	138	0.9612
19	1.0204	49	1.0055	79	0.9906	109	0.9756	139	0.9607
20	1.0199	50	1.0050	80	0.9901	110	0.9751	140	0.9602
21	1.0194	51	1.0045	81	0.9896	111	0.9746	141	0.9597
22	1.0189	52	1.0040	82	0.9891	112	0.9741	142	0.9592
23	1.0184	53	1.0035	83	0.9886	113	0.9736	143	0.9587
24	1.0179	54	1.0030	84	0.9881	114	0.9731	144	0.9582
25	1.0174	55	1.0025	85	0.9876	115	0.9726	145	0.9577
26	1.0169	56	1.0020	86	0.9871	116	0.9721	146	0.9572
27	1.0164	57	1.0015	87	0.9866	117	0.9717	147	0.9567
28	1.0159	58	1.0010	88	0.9861	118	0.9712	148	0.9562
29	1.0154	59	1.0005	89	0.9856	119	0.9707	149	0.9557

T = temperature (°F)
m = multiplier
The multiplier, *m*, will reduce the volume of fuel at any temperature *T* to its equivalent volume at 60°F.

APPENDIX A-3 CONVERSION BETWEEN VISCOSITIES AT THE SAME TEMPERATURES

Kinematic (cSt)	Saybolt Universal (s)	Saybolt Furol (s)	Redwood No. 1 (s)	Redwood No. 2 (s)	Engler (°)
20	98	*15*	86	—	2.87
22	106	*16*	93	—	3.10
24	115	*17*	101	—	3.34
26	124	*17*	109	—	3.58
28	132	*18*	116	—	3.82
30	141	*19*	124	—	4.07
32	150	*19*	132	—	4.32
34	159	*20*	140	*15*	4.57
36	168	*21*	148	*16*	4.82
38	177	*21*	156	*17*	5.08
40	186	*22*	164	*17*	5.34
42	195	*23*	172	*18*	5.59
44	204	*24*	180	*19*	5.84
46	213	*24*	188	*20*	6.10
48	223	*25*	197	*20*	6.36
50	232	26	205	*21*	6.63
55	255	28	225	*23*	7.24
60	278	30	245	*25*	7.90
65	301	33	266	*27*	8.55
70	324	35	286	*29*	9.21
75	347	37	306	*31*	9.87
80	370	40	327	*33*	10.5
85	393	42	347	*35*	11.2
90	417	44	367	*37*	11.8
95	440	46	388	*39*	12.5
100	463	48	408	*41*	13.2
110	509	53	449	*45*	14.5
120	555	58	490	*49*	15.8
130	602	62	531	*53*	17.1
140	648	67	571	*57*	18.4
150	694	72	612	*61*	19.7
160	741	76	653	*65*	21.0
170	787	81	694	*69*	22.4
180	833	85	735	*74*	23.7
190	880	90	776	*78*	25.0
200	926	95	816	*82*	26
220	1018	104	898	*90*	29
240	1111	113	980	*98*	32
260	1204	123	1061	106	34
280	1296	132	1143	114	37
300	1389	141	1225	122	39
320	1481	151	1306	131	42
340	1574	160	1388	139	45

APPENDIX A-3 (*Continued*).

Kinematic (cSt)	Saybolt Universal (s)	Saybolt Furol (s)	Redwood No. 1 (s)	Redwood No. 2 (s)	Engler (°)
360	1666	170	1470	147	47
380	1760	179	1551	155	50
400	1852	188	1633	163	53
420	1944	197	1714	171	55
440	2037	207	1796	180	58
460	2129	216	1877	188	61
480	2222	226	1960	196	63
500	2314	235	2041	204	66
550	2546	258	2245	225	72
600	2777	282	2449	245	79
650	3008	306	2653	265	86
700	3240	329	2857	286	92
750	3472	352	3062	306	99
800	3703	376	3266	327	105
850	3935	400	3470	347	112
900	4166	423	3674	367	118
950	4398	446	3878	388	125
1000	4629	470	4082	408	132
1100	5092	517	4490	449	145
1200	5555	564	4898	490	158
1300	6018	611	5307	531	171
1400	6481	658	5715	572	184
1500	6944	705	6123	612	197
1600	7406	752	6531	653	211
1700	7869	799	6939	694	224
1800	8332	846	7348	735	237
1900	8795	893	7756	776	250
2000	9258	940	8164	816	263

Index

W